Fur Trade Canoe Routes of Canada / Then and Now

Morse, Eric Wilton, 1904-
    Fur trade canoe routes of Canada; then
and now. [2d ed. rev.] Toronto, University of Toronto Press, in association with
Environment Canada, Parks Canada and the
Canadian Government Publishing Centre,
Supply and Services Canada, 1979.
    124 p. illus., map (Canadian University
paperbooks, 239)
    Bibliography: p. 111-113.
1. Fur trade - Canada. 2. Trade routes.
I. Title.
0802063845 pb   6/ex   0863912 LC

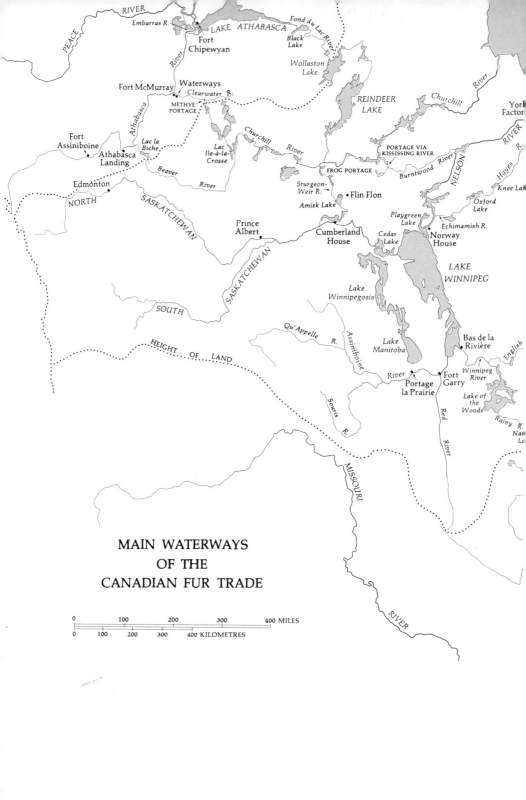

MAIN WATERWAYS
OF THE
CANADIAN FUR TRADE

University of Toronto Press
Toronto Buffalo London

Published in association with
Environment Canada, Parks Canada, and the
Canadian Government Publishing Centre,
Supply and Services Canada

Fur Trade
Canoe Routes
of Canada

Then and Now

ERIC W. MORSE

First published 1969
© Crown Copyrights reserved
Reprinted 1971
Second edition
© Minister of Supply and Services Canada 1979
Printed in Canada

Canadian Cataloguing in Publication Data

Morse, Eric W., 1904–
    Fur trade canoe routes of Canada

    Bibliography: p.
    Includes index.
    ISBN 0-8020-6384-5 pa.

    1. Fur trade—Canada. 2. Trade routes.
    I. Canada. Dept. of Indian Affairs and Northern Development.
    II. Canada. Parks Canada.
    III. Title.

    HD9944.c22M65 1979   382'.45'639110971   c79-094313-1

Cover illustration:
Canoes in a Fog, Lake Superior, 1869, by Frances Ann Hopkins.
Courtesy of The Glenbow Museum, Calgary, Alberta.

To all those, including my wife, who have
been my companions and fellow-voyageurs
on the rivers of the north.

Acknowledgements

I am indebted to Major-General N. E. Rodger of
Winnipeg, and to Dr. W. G. Mathers of Pinawa, Man.,
for important information as to the current state of
inundation of fur-trade rivers in their own Province.
I should like also to acknowledge the generous
co-operation of Major-General W. J. Megill and the
Royal Canadian Geographical Society in making
available maps and other material from a set
of articles I wrote previously in the
Canadian Geographical Journal.

E.W.M.
Ottawa
April, 1968.

Preface to Second Edition

Resulting from either terrain changes or new material
coming to light since 1968, three main changes in text
have been made in the Second Edition. They relate
respectively to Great Dog and Rat Portages and to
Boundary Falls (Portage de l'Isle), all between pp.
76–86. An index also has been added.

Frances Hopkins's contemporary, almost "photographic" painting, "Bivouac of a Canoe Party" shows various evening camping activities of a small brigade of two North Canoes in progress, including wood-gathering, cooking, insect-defence, and sleeping. At the right, side by side, are two different paddles, as used by the *milieux* and the *bouts*.

Contents

ILLUSTRATIONS

courtesy of Public Archives of Canada, Ottawa:
pages viii (C-2773), 6 (PA-59510), 9 (C-2772), 18 (C-16442), 28 (C-6690),
41 (C-7960), 49 (C-2390), 58 (C-13585), and 104 (C-70270);
courtesy of The Glenbow Museum, Calgary, Alberta:
cover and pages 10–11, 45, and 82;
Eric Morse: page 51;
B. M. Litteljohn: page 72;
Parks Canada, Department of Indian and Northern Affairs: page 103.

Foreword

It has taken an unusual combination of qualities to produce this book. Everyone who has canoed with Eric Morse knows of his capacity for endurance, his skill with a paddle and his enjoyment of "white water". To this he has added years of research, both in the library and on the river. In the best tradition of Francis Parkman, he has made himself familiar with almost every mile of even the most rugged and remote of the voyageurs' routes. The result has been exciting adventure, much scholarly recreation and this vivid contribution to our history.

In the past the teaching of history in our schools has been dominated by traditions inherited from Europe. On that continent history has been filled with battles, and the lives of national heroes. In Canada we have had few decisive battles and not many dominant leaders. Much more important to our history has been the struggle of nameless Canadians to improve their lives in our often hostile environment. This struggle has produced its share of adventure and heroism.

But perhaps this lesson is best learned outside the classroom. Anyone who wishes to get a feeling for the unique history and geography of this country can do no better than follow Eric Morse's example. All who do will derive a double pleasure from this book: the pleasure of reading it and the pleasure of exploring some of the routes it describes. Like those of us who have canoed with the author, I am sure they will come to share his enthusiasm for the tough but satisfying life of the voyageurs.

Rt. Hon. Pierre E. Trudeau
PRIME MINISTER
Ottawa, 1968

Introduction

In its simplest economic terms the Canadian fur trade was the marriage of a demand in Europe for a particular fur, and the demand in North America among primitive tribes for certain products of Europe's new and expanding industry. The most sought-after fur was beaver, in the ruthless pursuit of which the fur traders had to go farther and farther west and north. When, from 1778, they reached the frigid Athabasca and Mackenzie regions, the traders found a quality of fur which made the depleted and warmer areas to the south and east relatively uneconomic to trap. A transport and supply system spanning 4000 miles had to be developed. Overcoming the formidable physical difficulties in moving the furs through a million square miles of northern wilderness out to the sea at Montreal or York Factory, and carrying back the trade goods — these essentially comprise the story of the fur trade in Canada. Like many other exciting Canadian stories, it was one of meeting the challenge of geography; and like many other Canadian challenges, it had to do with transportation. In the absence of road or rail, everything had to be transported by rivers and lakes, where the most suitable craft, already developed by the natives, was the canoe.

Much has been written on the *history* of the Canadian fur trade, but less on its geography. The purpose of the present volume is to define the fur-trade canoe routes, linking them where necessary with modern landmarks and roads, to describe their general condition today where they have been changed, and indirectly to stimulate interest in visiting and preserving them.

This study is a by-product of several years' summer holidays devoted to retracing by canoe the principal routes, with early journals excerpted and taken along for study on the spot. These routes in the nineteen-sixties remain dramatically unchanged from what the first fur-seekers saw three centuries earlier. But hydro, industry and settlement obliterate and threaten; and some day it might be too late to undertake as close an examination as we are happily still able to make.

The plan of the book is, in Part I, to introduce the subject as seen through the eyes of the real hero of the story, the Canadien (or Indian, or

1

Orkneyman) voyageur, to outline the economic and geographical influences and then, in Part II, to describe in detail the old routes as they look now. Original spelling and punctuation have been preserved in all quotations from early journals. A few coded references relate to the bibliography and offer suggestions for further reading.

PART ONE

Chapter I

The Voyageurs On The Trail

"They preferred running the Dalles; they had not gone far, when to avoid the ridge of waves, which they ought to have kept, they took the apparent smooth water, were drawn into a whirlpool, which wheeled them around into its Vortex; the Canoe with the Men clinging to it, went down end foremost, and [they] all were drowned; at the foot of the Dalles search was made for their bodies, but only one Man was found, his body much mangled by the Rocks." — David Thompson's Narrative.

The trade routes were rivers and lakes — and not necessarily peaceful reedy rivers, nor pleasant mirrored lakes. They took their toll of lives. When a voyageur died on the trail, his companions buried him and over his grave erected a crude wooden cross. The early journals speak of clusters of crosses along the way — sometimes as many as twenty or thirty in one spot. The commonest causes of fatality appear to have been drowning, strangulated hernia, or injury in a fall. The commonest locations of tragedy were rapids and portages.

The fur-trade period has been depicted as Canada's epic age, which may be true. But the glamour should be muted, and considered against the toil of voyageurs working sixteen and eighteen hours a day, often immersed for long periods in icy water or carrying 180-lb. loads through bog, up hills, and over long portages. Other hazards haunted the voyageur as he worked a canoe containing three tons of cargo up a foaming rapid, or navigated the canoe down some roaring canyon. Or, as he coasted around an inland sea like Superior, perhaps suddenly half-way across a great bay a squall struck. Squall, toil, hazard and hardship dogged the voyageur on the trail.

Yet to face these he had impressive skill. What were the techniques he employed? Going upstream, when the current became too fast to paddle

Photo of a party poling up rapids on the Abitibi River. The second canoe is a birch-bark, roughly modelled on the North Canoe. The other is a modern canvas "Prospector" model. The poles used are 8 to 10 feet, with a metal ferrule.

against and yet not strong enough to require leaving the water, there were two means of attack: provided that the shoreline was reasonably free of snags, the canoe was lined (tracked); otherwise it might be poled. In lining, the voyageurs attached themselves like canal horses, sometimes by a one-shoulder harness, to a line 60 to 100 feet long, while the steersman stayed in the canoe. To guard against the risk of the towline breaking, the line (*cordelle*) consisted of half a dozen smaller lines braided together. Poling was accomplished with a long, eight or ten foot pole (*perche*), usually shod with metal. The polers stood, and the technique called for perfection in timing and balance. The river bottom for poling had to be fairly shallow and firm; under ideal conditions a quite considerable rapid could be ascended in this way without the toil of portaging. Where the shoreline was too cluttered for lining, the bottom too deep or soft for poling, and the current too swift for paddling a fully-loaded canoe, there was yet another technique, which consisted of taking the rapid in two bites, *demi-chargé*. Half the packs were taken out, the rapid was forced, the half-load deposited, and the voyageurs then went back for the second half-load.

But a strong rapid required sterner measures. A ridge, a fall, or a shallow rocky area in the rapid meant that the whole cargo had to be portaged. If the empty canoe could then be hauled up the rapid, at least those voyageurs responsible for carrying it had this much respite; this was called a *décharge*. A portage was a last and inescapable extremity. With a view to the portaging, all packs whether of furs or trade goods were made to weigh roughly ninety pounds. Because of the number of packs carried in a canoe, it worked out that each voyageur became responsible for portag-ing about half a dozen packs (called *pièces*), in three loads. A pack was put in a leather sling or tump-line which went above a man's forehead; into the hollow thus formed behind his neck he tossed another 90-lb. pack. With this 180 pounds — or more if he was showing off — he half-ran at a slightly bent, shuffling trot. If the portage was more than a half a mile (a ten-minute carry), the voyageur, in order better to distribute his loaded and unloaded periods, dropped his packs at what was known as a *pose* and went back for the next load. *Poses* were about half a mile apart.

The method of carrying the canoe depended on which of two types of canoe was to be portaged. A 36-foot *canot de maître* weighed about 600 lb. It was portaged by four men supporting the gunwale on a padded shoulder — two men near the bow and two near the stern — in the ordi-nary (inverted) way a canoe is portaged. West of Lake Superior the North

Canoe was used: about 25 feet long and weighing about 300 lb. This required only two men in portaging, unless the portage were specially steep or long. Owing to its shape, and in order to allow proper freedom for vision and leg-movement a North Canoe was portaged in its *upright* position, as it would sit on the water, the men carrying it on the shoulder.

Though the average speed for a man portaging was three m.p.h., with his three loads over (and two trips back, unloaded) a voyageur would have to walk a total of five miles for every mile-long portage. Including time for loading and unloading, a mile portage thus would take a total of two hours — or only half a mile an hour, net. All of which helps to account for occasional instances where a more roundabout water route was selected by the freight canoes, in preference to the shorter Indian canoe route involving longer portaging.

Coming downstream, the difficulties were differently handled. A shallow portion of a stream could be lined past; or the men would simply wade, guiding the canoe. An unnavigable rapid or a waterfall would dictate a portage. Landing at a portage in fast current called for a little more dexterity, especially in not letting the birch craft touch even a root. The men jumped into the water up to their waists, and the canoe was unloaded in the water.

The least strenuous and most exciting interludes in the voyageurs' life were when they came to navigable rapids. A mile or so away, as they came down some big river like the Churchill, the rapid would first catch their ear; then around a bend in the river the first dancing spray was seen. The canoes in the brigade would pull into an eddy and hold back while the *guide* in charge inspected to see whether the present water-level permitted this or that course. To the *guide*, the rapid would be an old friend — or enemy; he had known it over the years as he had risen in the ranks from a young, inexperienced *milieu*. He might not always need to inspect it. Then the signal would be given and, spaced a little apart, the brigade would enter the rapid, line astern, and become committed. In a big canoe there is not much that the middle men can do except, at the orders of the *avant*, in charge, to increase the speed of the craft or to brake it. Unlike a two-man canoe, a freight canoe cannot pivot, nor be pulled sideways. It cannot navigate a "slalom" course. The stern man, the *gouvernail*, standing up with his nine-foot paddle, had the greatest purchase in steering; but he was limited in vision by the size of his vessel. Thus the key man in shooting rapids was the *avant*, who also stood; he

This Frances Hopkins oil painting is inaccurately entitled, "View of a canoe party round a camp fire". Voyageurs had no time for sitting around a camp fire. This detail from the painting, in fact, shows some skilled canoe-repair in progress; the fire is for boiling pitch.

Frances Hopkins's well-known oil painting, now in the possession of the Glenbow Museum, Calgary, shows *canots de maître* paddling in the fog on Lake Superior.

The end-men are standing. The *gouvernail* in the nearest canoe indicates by his grip on the paddle that he is steering, not paddling. Note the tip of the dismantled mast.

too had an out-sized paddle with which to pry or pull the bow around.

The rapid would probably start off in a long, narrowing, satin-slick "V", down which the *milieux* would paddle hard to give the canoe steerage. At the foot of the "V" would begin a train of high, standing waves. The canoe course in a shallow rapid would lie through the biggest waves; in a deep, turbulent rapid the canoe would be headed just outside the standing waves, yet safely away from the rocks and ripping eddies nearer shore. Piercing rocks below, swamping waves above, were the twin dangers —skirted past in a craft made of birch rind, with a freeboard of six inches.

The voyageur's daily routine was a killing one: for the six-to-eight weeks he was on the road, he was roused as early as two or three a.m., and (provided that no rapids were just ahead) he set off breakfastless. Before eight o'clock a breakfast stop was made. A midday lunch was "served" — though often lunch was no more than an opportunity to hack off a piece of pemmican to chew on the way. However, fairly regularly, a stop was made for a few minutes each hour to allow the men to have a pipe. This event was so important that distances came to be measured in *pipes*; "*trois pipes*" might be 15 to 20 miles, depending on wind and current.

Between eight and ten in the evening, depending on the light, camp was made. Supper, which might have been pre-cooked the night before, was warmed up and served. The men then dropped down on turf, moss, or beach, with their heads under the overturned canoe (whose high end lent itself to this purpose). A tarpaulin would be stretched from the canoe to give shelter from rain and dew. There was no provision ordinarily made for protection against insects, except the voyageurs' own grease and dirt, assisted sometimes by a smudge, lit up-wind.

The voyageur employed by the North West Company was the Canadien — tough, gay and carefree. Without him the Montreal fur trade could obviously not have been carried on. He was incredibly durable, and also tractable. The voyageur is not to be confused with the *coureur de bois*, who, though he too paddled and carried, was an independent trader, not a hired labourer. There is, moreover, a time-distinction: by and large, the *coureur de bois* belongs to the French period of Canadian history, while the voyageur belongs to the period after 1763. Nearly every journal had a description of voyageurs. Many of these early accounts were written, perhaps a little lyrically, by just a passenger in the canoe. The following contemporary description was written by the fur trader, Daniel Harmon, in the early nineteenth century. Having employed them and lived with them for years, often in conditions of danger and adversity, Harmon knew voyageurs intimately; and he presents on the whole an admiring picture, yet one displaying all the warts and pimples:

> *"The Canadian Voyagers possess lively and fickle dispositions; and they are rarely subject to depression of spirits, of long continuance, even when in circumstances the most adverse. Although what they consider good eating and drinking constitutes their chief good, yet, when necessity compels them to it, they submit to great privation and hardship, not only without complaining, but even with cheerfulness and gaiety. They are very talkative, and extremely thoughtless, and make many resolutions, which are almost as soon broken as formed. They never think of providing for future wants. . . . They are not brave; but when they apprehend little danger, they will often, as they say, play the man. They are very deceitful, are exceedingly smooth and polite, and are even gross flatterers to the face of a person, whom they will basely slander, behind his back. . . . A secret they cannot keep. They rarely feel gratitude, though they are often generous. They are obedient, but not faithful servants. By flattering their vanity, of which they have not a little, they may be persuaded to undertake the most difficult enterprises."*

John Jacob Astor said he would prefer to have one Canadien voyageur to any three others.

Quite as heroic a figure as the Canadien voyageur — though perhaps less "romantic" — is the Hudson's Bay Company Orkneyman voyageur. His sufferings, his toughness, were no less: indeed in a way he is even more to be admired, for he was contending with new conditions in an alien land.

The voyageur, whatever his origin, was the motive power of this transportation system. To aid paddling under conditions of difficulty or monotony, the Canadien voyageurs had songs, which they sang lustily at the paddle. These songs — "En Roulant ma Boule", "C'est l'Aviron", "A la Claire Fontaine", Youpe Youpe sur la Rivière", etc. — gave to their stroke rhythm and drive, performing in a way the function of sea shanties for sailors.

"They were hurled down with surprizing velocity thro' three suc-

cessive cascades. . . . The canoe was several times overwhelmed with water, and threatened every moment with being dashed to pieces in the windings of the Rocks . . . and it remained a considerable time under water: — At length however the Current drove it toward shore, with the [two] men still hanging after it, and tho' they at first seemed unsensible, yet . . . they recovered their strength and before night renewed their labours with as much alacrity as if nothing had happened to them." — Journal of Duncan M'Gillivray.

Chapter II

The Economics And Logistics Of The Trade

The Economic Drive

By far the most important fur was *beaver*, sought not for the warmth of its pelt but for a unique feature of its hair: it is minutely barbed. This helps the animal to trap a layer of air against its body for thermal purposes, especially useful in the water. The little barbs are on the beaver's "wool" (or *duvet*), underneath the longer, glossy guard hairs. In preparing beaver for commercial use, the guard hairs were first removed by one process or another, and the wool was scraped off the pelt. Skins which had been worn by an Indian as clothing (*castor gras*) were most in demand, because the guard hairs had already worn off.

Beaver was used in the felt industry. The hair was chopped up and mixed with cheaper hair such as rabbit or cat. Under the pressure of the felting process, the barbs served to bind the whole together and made for strength and quality. A beaver hat was the hat of fashion, and bore no resemblance to a frontier coonskin; the felt, stiffened with shellac, was moulded into top hat, tricorne, or flat shape, according to the vogue of the day.

Though beaver dominated the trade from the start, it was far from being the only fur trapped. Milady accepted — or purchased — mink, marten, fisher, otter, ermine, fox, lynx; while coarser skins such as bear and wolf also found a market. The trade also carried deer, moose, and buffalo skins, and even the skins of wild swans and geese.

By the sixteenth century Russian and Baltic fur sources were becoming depleted. Under these conditions of both strong demand and world supply, the Canadian fur market held firm from the earliest contact with natives until past the middle of the nineteenth century, when the new fashion of silk hats ruined the beaver market. Sustained demand often led, through victimization and debauching of primitive tribes, to fantastic profits of 1000% and 2000% — though war, cut-throat competition, losses at sea, and a three-to-six-year lapse between investment and return, demanded high compensation. The vast transportation system described below could have been supported only by a commodity of such concentrated value.

Apart from the demand for furs, there was at the same time a strong demand among Indian tribes emerging from the Stone and Copper Ages, for the white man's manufactured goods. Once one tribe had acquired firearms, it became a matter of simple survival for its neighbours to seek the same. Bows and arrows were relegated to killing game when ammunition was short. In hunting, the Indian had managed to live in a state of equilibrium with his game supply; and his primitive hunting tools took care of this. Once the fur trader's commercial greed and goods entered the picture, things changed. Instead of taking a beaver or two from a lodge, the whole colony was taken; and for this, more efficient killing implements were needed: traps, ice chisels, axes, knives, hooks, guns and ammunition. In the matter of clothing and bedding, the traders made the Indians' life more comfortable and secure. The Indian soon became dependent on the traders' blanket coats and woollen trousers and socks, and also on blankets.

Indian women, in a life of threadbare slavery, benefitted perhaps most from the trade goods. Apart from mirrors, beads and trinkets, they prized instruments for cooking and sewing. Till the advent of metal cooking vessels, the process of boiling was performed by dropping red-hot stones into vessels of skin or birchbark. Before the needle, only a bone-splinter or a thorn was available for sewing; while sinew or rootlet had served for thread.

Such then was the reciprocal drive which powered the fur trade. From here, this book is concerned less with what was traded than with where and how the goods were carried.

River-mouth Trading

The first trading contacts between Europeans and the Indians inhabiting what is now Canada probably took place in the general area of the Gulf of St. Lawrence. The Europeans were fishermen landing to dry their fish. The Indians traded the beaver robes they stood in, for whatever in the way of knives and needles the fishermen happened to have. In this first fur-bartering, formal trade routes were not required: the Indian came to rendezvous with the fishing fleets at the proper season. The Maritimes area was so cut up by arms of the sea that the Indian rarely had a great distance to carry his furs, even after the trade became more organized. Since the Indian canoe could not put out to sea, nor the sea-going fishing craft venture far up a stream, river-mouth trading became in this area almost exclusively the rule. The Saguenay was particularly important in this regard; in the Maritime Provinces almost any river near the fishing boats served.

River-mouth trading, too, was for its first century the general practice of the Hudson's Bay Company.

Coureurs de Bois and Indian Middlemen

As the fur trade of New France, on the other hand, gradually extended itself, the traders began to push inland up the rivers. During an interlude until the incursions of the Iroquois ended the practice, the Indians were encouraged to bring their furs down to the great fur fair on the Island of Montreal. More and more the *coureur de bois* took his canoe-load of trade goods out to the Indian, lived with him, and as he lived, returning in the spring with a cargo of furs. Even the early fur companies such as the Company of the Hundred Associates managed to operate along these loosely organized lines. Trade routes, such as were used later, had hardly emerged; any river or creek served the *coureur de bois* as a "trade route". Nor did he yet need to go far in pursuit of his pelts.

Another factor minimized the need for trade routes at this stage: this was the Indian middleman, operating both with the Montreal traders and with the Hudson's Bay Company. Bartering for their trading goods at the posts, the middlemen traded these, at an appropriate mark-up, to tribes farther inland. With the Hudson's Bay Company it was the "Home Guard" Indians, the Swampy Crees, who served as middlemen. Among the Indians of the St. Lawrence and Ottawa, the Ottawas, the Nipissings, and the Hurons played a leading role in this respect.

River-mouth trading, the winter sorties of the *coureurs de bois,* and the work of the Indian middlemen broadly took care of the transportation of furs till around the end of the seventeenth century. About then Kelsey proceeded to the Prairies for the Hudson's Bay Company. When shortly afterward La Vérendrye did so for the Montrealers, the trunk routes began stretching out and shaping up — the Ottawa, the Great Lakes, the border-lakes route west from Grand Portage.

This old photograph shows a H.B.C. freight canoe being portaged as a *canot de maître* (or Montreal Canoe), inverted, with the gunwale on the men's shoulders. The *canot de maître* was portaged by four voyageurs.

New Sources of Beaver

Two new factors emerged about now to give the fur trade the beginnings of its continental dimensions. These have been alluded to in the Introduction: beaver extermination and the discovery of new standards of fur quality, the result of colder climate, in the Northwest.

The beaver had no defences but its "moat", and its teeth at close quarters. In David Thompson's words, the Indians

> "procured from the French Axes, Chissels, Knives, Spears and other articles of iron . . . Thus armed, the houses of the beavers were pierced through, the Dams cut through, and the water of the Ponds lowered, or wholly run off, and the houses of the Beaver and their Borrows laid dry."

By the 1750's there were already signs of depletion on the lower Ottawa. Farther and farther west, ruthlessly and relentlessly, the beaver was trapped. In the process of pushing westward and northward for new sources, Peter Pond in 1778 made the epochal first crossing of the Methye Portage by a white trader. He wintered on the Athabasca River, forty miles south of Athabasca Lake, and cornered so many choice furs that he was forced to leave half of them behind for another trip. Fellow Montrealers followed him over the Methye Portage; and it was not long before Fort Chipewyan was built on Lake Athabasca, and soon moved to the north shore. The Athabasca Country became the El Dorado of the Canadian fur trade, abounding in small streams and in poplar, lacking in settlement, and above all very *cold*. The Athabasca beaver required a very much better fur coat than did his cousin wintering in Virginia or on the Gulf of Mexico. A further advantage for the Montrealers was that the Athabasca Country lay in the Mackenzie basin, and therefore technically outside the writ of the Hudson's Bay Company charter. Not that there was great evidence of this being important in the Montrealers' eyes — nor did it deter the Hudson's Bay Company from following them over the Methye Portage later.

Organized Transportation Begins

The sheer distance of the Athabasca Country from Montreal demanded better co-ordination of transport. At the same time, from about 1778, various independent Montreal fur traders had begun to merge into loose trading alliances; and in 1784 the North West Company was formed on a more corporate basis. The stage was now set for a highly organized transportation system. Another important and concurrent development was a change in the policy of the Hudson's Bay Company, as it departed from its river-mouth trading on the shores of Hudson Bay, and came inland to meet the Montrealers' competition.

The logistical problems of the Hudson's Bay company were minor at this stage. A trunk line only 350 miles long connected York Factory, at the mouth of the Nelson-Hayes estuary, with Norway House at the north end of Lake Winnipeg. From Norway House two main routes went, one 250 miles to Cumberland House, and the other 300 miles to what is today Winnipeg. Later, after the absorption of the North West Company in 1821, the Hudson's Bay Company's problems became vaster in area and more complicated in character.

The North West Company in its last forty years, 1780 to 1820, had to overcome four formidable physical problems in transport:
— the Athabasca Country was 3000 miles from Montreal;
— there were only five months from break-up to freeze-up;
— a freight canoe could at best average only 1000 miles per month;
— the canoe required for Lakes Superior and Huron was too big for the smaller streams and portages of the country to the west of Lake Superior, and *vice versa*.

The Montrealers solved these problems by having two sets of crews and canoes. A great central *entrepôt* was established at Grand Portage, which was shifted about 1803 to Fort William. The two brigades of canoes set out from opposite ends, Lachine and Fort Chipewyan, respectively in early and mid-May, and met at the *entrepôt*, where they exchanged cargoes. A month was used for "turn-around", to recuperate the crews, to repackage the goods, and to arrange consignments for various posts along the way. By the end of July, the two sets of voyageurs were on their way back again. In this way, furs reached the east, and trade goods the west, within the five-months frost-free span. The Athabasca Brigade was met at Rainy Lake by a special brigade from Grand Portage and made their exchange of loads there, in order to get a head start home.

The Contribution of the Indian

No proper appreciation can be gained of where the canoe routes lay, nor why, without knowing the capabilities of the craft the fur traders used, nor the "fuel" which the human engines burned in crossing a farm-less wilderness. Before discussing the craft and the provisioning, a tribute should be paid to the Canadian Indian. The Indian showed the trader *how:* the fur canoe was the ordinary Indian birchbark canoe, but developed to its maximum strength and capacity; the provisions for the voyageurs while *en route* were almost entirely Indian in origin. Indian methods of fishing through the ice helped the trading posts through the winter; for travelling in winter, the trader adopted the Indian snowshoe, toboggan, and dog team. The Indian, moreover, showed the trader *where,* for the fur-trade canoe routes were not discovered and engineered by white explorers: some had been in use since the passing of the last ice age. The Indian guided the explorer in.

The Friendly Indian may be a new concept for people reared on Fenimore Cooper's *Leatherstocking Tales,* or on the American frontier philosophy of "The only good Indian is a dead Indian". The fact is that the economies of Canada and the U.S.A. were essentially opposite in character during most of the fur-trade era: American settlers were pushing westward tilling the land, land which had to be taken from the Indian — who thus, because of his understandable reaction, became equated with bears and wolves, to be exterminated for safety. As compared with the American thrust for settlement, Canada's was still a wilderness economy, based on *partnership* with the Indian, who in turn needed the traders' goods. A special rapport, in fact, existed between the Indian and a good fur trader — cemented by many a casual alliance with a chief's daughter while wintering at a trading post. The Hudson's Bay Company, by generally fair dealing, held the Indians' respect. This was a powerful influence in keeping peace in the Northwest, before the forming of the Northwest Mounted Police. Not that there were no incidents nor bloodshed — even massacres — north of the border; most of these, however, are traceable to tribes of Indians who inhabited the Prairies and whom the buffalo made relatively independent of the trader; or similarly to tribes along the Pacific Coast reared to independence and a high degree of civilization by the security and plenty of their salmon economy. Quite obviously the Canadian fur trade, regularly plying the broad emptiness between the St. Lawrence and the Rockies, could not have been carried on without basically friendly relations with the Indian tribes along the way. Indians caused no serious interruption of trade in Canada, after the pacification of the Iroquois in the late seventeenth century.

The Canoes in the Fur Trade

There were two basic canoes which were used in the fur trade: the *canot de maître* (or Montreal Canoe) and the *canot du nord* (or North Canoe), the main difference being their dimensions. Between them was a hybrid type, the *canot bâtard*. There is no need here to go into details of canoe construction nor tribal variation; the classic and definitive work on this is Adney and Chapelle: *The Bark Canoes and Skin Boats of North America* (Smithsonian, Washington, 1964). For practical purposes of understanding the fur trade routes, the features to note are such things as *capacity, freeboard, size of crew,* and the particular *advantages* and *disadvantages* of special craft.

The *canot de maître* extended the Indian birchbark canoe to its ultimate as a cargo vessel. It was basically thirty-six feet in length (extreme cases of forty feet are recorded), six feet in beam; it carried a fantastic payload of three tons, plus its crew and their personal gear — totalling four tons gross. This, in a craft lacking an ounce of metal. The *canot de maître* was big enough to cope with the waves and rapids of the Great Lakes and the Ottawa; it was too big, however, to navigate some of the smaller waters lying near the two continental divides to be crossed between Lake Superior and Lake Athabasca. The North Canoe, used for this run, had just half the capacity — a ton and a half of cargo, 3000 lb. The North Canoe's basic dimensions were 25 feet in length, four to four and a half feet in beam; it had, fully loaded, only about eighteen inches draft. Specifically, the Montreal Canoe carried sixty-five 90-lb. *pièces;* the North Canoe, 25-30 *pièces.* The canoe's severe limitation on cargo is reflected in a voyageur's personal baggage-limit of 40 lb. (interestingly identical with the earlier limits on aircraft). More significantly, the limitation affected provisioning en route — which in turn sometimes determined where a fur-trade route could go.

The maximum length of a freight canoe was a function of the stress which a keel-less bark craft could take, combined with whether it was designed for big lakes or small rivers. Given this limitation on length, the determining factor in the number of packs a canoe could carry was safety — how far down in the water the canoe sat. The freeboard aimed for even in a Montreal canoe appears to have been as little as six inches. The low freeboard of a freight canoe is the proper context perhaps in which to refer to the paddle the voyageurs used. While the *avant* and *gouvernail*, at bow and stern, required a paddle of seven to nine feet for "purchase" when standing, the *milieux*, sitting close to the water line, required only very short paddles, little more than four feet. The ordinary voyageur's paddle, moreover, was narrow and stocky, for use sometimes in fending in a rapid. A favourite wood was basswood *(bois blanc)*, stronger than spruce or pine, lighter than birch or maple.

There are authors who seem inclined to argue for a set number as crew; but in reading all the various journals, I have found little to support the idea of uniformity. *Canots de maître* are recorded as setting out with crews of as few as six, and as many as twelve. These numbers, of course, are extreme: six would lack power; twelve would be at the sacrifice of cargo. Eight to ten appears to have been the average crew of a Montreal Canoe. Comparably, there is a spread of between four and eight recorded as the crew of a North Canoe, with a common average of five or six. Factors which made for an extra small, or an extra large, crew were: new voyageurs going in, or term-expired "winterers" going out working their way; difficulties to be encountered *en route*, as over the Rockies; shortage or sickness of personnel at a post. In sum, there is evidence of considerable spread and flexibility in size of canoe crews, the biggest crew of a North Canoe sometimes even exceeding the smallest crew of a Montreal Canoe.

The above crews applied to freight canoes. There was also the *express* canoe, bearing orders, mail, or news of a shift in the market, or of war. These were the fur companies' only means of communication, and the stress was not on capacity but on *speed*. Here a crew of fourteen might be assigned to a Montreal Canoe. For an express canoe the *canot bâtard* was sometimes used; its compromising dimensions allowed it to travel anywhere. A North Canoe, if used for express purposes, might carry a crew of eight or nine.

The freight canoe was modified from a primitive craft which over the years had evolved to meet terrain requirements within the limitations of available material. In a terrain whose waterways were interrupted by chutes and unnavigable rapids, the need was for easy portability; in a country abounding in birch, the materials for making and repairing canoes were constantly at hand — including *wattape* (spruce roots) for sewing the bark sheets together, spruce and pine gum for caulking, hardwood for ribbing, and split cedar as flooring to spread the load evenly. Each canoe on the road carried its "spare parts" in a roll. In the rough wilderness conditions in most of Canada, nothing has yet been developed to beat the canoe. However, a birch canoe lacked strength and durability, the life expectancy of a freight canoe being little more than a year. Still another limitation was the skill required by the crew, calling for a pool of experienced voyageurs to draw from.

It was such restrictions which persuaded the Hudson's Bay Company, more particularly in the regime of George Simpson, to abandon the use of canoes on trunk routes from Lake Winnipeg, north-east to York Factory, south to Fort Garry, and west to Edmonton. Instead, the Company set about employing a boat already developed for use on the Albany River, which came to be known as the York Boat. Fortunately, one of these has been preserved at Lower Fort Garry, near Winnipeg. It carried few more *pièces* than a Montreal canoe, but had a smaller crew of only six to nine (which meant more cargo in proportion to crew), was more seaworthy on big lakes, was capable of carrying more sail in high wind, could be *rowed* by Orkneymen unskilled at the paddle, and was much more durable. The York Boat, with pointed bow and stern, was a little larger than a Montreal Canoe. The Hudson's Bay Company claimed that York Boats could, in terms of wages and durability, be operated and maintained at a 33% saving over canoes. The York Boat's limitation was, of course, its lack of portageability. Too heavy to be carried, it had to be dragged or winched on rollers over portages. This was feasible on rivers in the Prairies and Hudson Bay areas, but was a handicap on rivers in the Precambrian Shield, such as the Churchill or middle Hayes. Rivers where York Boats took over from canoes will be treated for purposes of this book as being canoe routes, the difference being largely one of time.

Some references (First number refers to bibliography)

3	I 131-2	20	13-15	29	5, 17-18
9	II 491, 629-30, 871-6	23	33-4	33	345-6
14	93-4, 118, 150	27	35, 53-4, 67	34	287-8, 293, 310, 382
18	1-2, 13-14	28	313-4	41	193, 207, 213-28

Provisioning on the Road

"Fuel" supply on the road was a major problem for the fur companies. Men using 4000-5000 calories a day had to be fed as they crossed a wilderness; they had time neither to hunt nor fish, in the explorer's sense of living off the land. A critical question was how much and what food could be carried?

The easiest portion of the Montreal main route to provision was that starting off at Lachine, in farm land near a sea port. Here the staple diet was dried peas or beans, sea biscuit and salt pork. This was considered in the fur trade as a fairly lush diet; and the voyageurs plying the run between Montreal and Lake Superior were known derisively among the tougher breed wintering in the North West as "pork eaters", *mangeurs de lard*. The fur-trade staples were in a sense prototypes of modern dehydrated foods, making it possible for enough provisions to be carried from Lachine all the way to Michilimackinac or Sault Ste. Marie, a distance of 750 miles. Such dry foods as beans, corn, and peas required reconstituting and pre-cooking the night before, making the cook's day extra rough and long.

The area about Lake Huron was maize country, another Indian commodity. Indian corn was made into a sort of hominy or corn mush, with suet, bacon fat or bear grease added, the Michigan counterpart of pork and beans. The Indians brought corn in for trade, from as far away as Detroit.

Between Sault Ste. Marie and Lake Winnipeg, wild rice and maple sugar could often be picked up *en route;* but the *basic* provisions for this 1000-mile section had to be taken in by the canoes as they travelled from east or west, for the wild rice (actually an oat) was not sustaining enough, nor quite reliable in supply. Sailboats on Lake Superior later assisted in provisioning Grand Portage.

The great food staple of the Western areas was pemmican, used from Rainy Lake to the Rockies. Pemmican was sun-dried, pounded buffalo (or it could be caribou or moose) meat, done up in 90-lb. hide containers with melted tallow poured over it. "Quality" pemmican had saskatoon berries mixed in. This was a very concentrated, highly nutritious food, three or four pounds being all that a slaving man would require as his daily ration. Kept reasonably dry, it lasted for months. The pemmican was either sliced off and munched raw, or made with flour and water into a *potage* called "rubbaboo". The canoes' limitation on cargo made necessary certain refuelling points. A North Canoe with six men already carrying 25 *pièces* would need to take aboard four *pièces* of pemmican to "fuel" it for the next 500 miles. This distance was the general spacing between the three North West Company pemmican posts: *Bas de la Rivière* (later Fort Alexander) at the

25

mouth of the Winnipeg River, using pemmican collected in the Red and Assiniboine valleys; *Cumberland House,* using the pemmican of the Saskatchewan valley; and *Ile-à-la-Crosse,* the outlet for Beaver Valley pemmican. The Hudson's Bay Company's chief pemmican posts were Fort Garry, Norway House, Cumberland House, and Edmonton.

Whitefish and salmon supplemented pemmican in supplying voyageurs' needs west of Ile-à-la-Crosse and Edmonton.

Some References

9	I 8; II 539	**18**	19, 111, 151-4	**32**	4-5, 20, 31-2
14	151, 197-8	**20**	16, 54, 184, 213, 275	**34**	218, 314-5
15	68, 108	**27**	35, 53, 94-96	**43**	342, 347-8, 451-4, 698

Chapter III

The Influence Of Geography

The climate, the soil, the amount of fresh water surface, the drainage pattern, and the Precambrian Shield — in a word, *geography* — dominate Canadian fur-trade history.

The two main contributions of Canadian climate to the fur trade, as already suggested, are in the *quality* of the furs and in the *haste* required in moving the goods so far through northern waters in the ice-free months. The haste in turn determined logistics: the central *entrepôt*, the two sets of canoes, a provisioning system for crews moving too fast to live off the land.

The really amazing feature of Canadian geography as it affects this story is the amount of fresh water in Canada: half of the whole world's fresh water surface. The rivers are not only closely connected but, for a portable craft adapted to the interrupted navigation, are entirely navigable. Even today it is possible to travel by canoe from almost any city in Canada to the Atlantic or the Pacific, or to the Arctic Sea.

Such flexibility of movement is not the result of canals, as in Europe or Asia, but of a natural drainage pattern consisting of three vast basins having very low rims. The whole area east of the Rockies, comprising three-quarters of Canada, presents no serious barrier to canoe travel. Ten per cent of Canada's surface drains out by the St. Lawrence; twenty-two per cent by the Mackenzie; and no less than forty-three per cent drains into Hudson Bay. The other twenty-five percent of the surface consists of all of the Pacific slope and a small part of the Arctic slope, neither of which bore trunk routes for the heavy carriage of furs.

Between the three great basins, the connecting portages on the trade routes were mostly short, the only two long ones being the nine-mile Grand Portage out of Lake Superior and the twelve-mile Methye Portage into the Mackenzie basin. No main portage was much over 1500 feet in altitude; the biggest single drop was 700 feet, at the north end of the Methye Portage.

Canada's almost total navigability by canoe is related also to the fact

For lack of any painting or photograph showing canoes being lined (tracked) this turn-of-the-century photograph is included to show barges being lined up the Athabasca. Each barge has seven men on the tracking line. A canoe would require far less. Usually the crew was divided into two alternating shifts.

that half of its surface lies on the world's oldest land mass, the Precambrian Shield, whose peaks and precipices have in billions of years been ground to gentler gradients. The great contribution of the Shield in making throughways is the chain of vast, connected lakes which it caused to be formed around its edge. Great Bear, Great Slave, Athabasca, Winnipeg, Lake of the Woods, Superior and Huron all have one side, or end, in granite. As the rains and ice-melt for thousands of years poured down from the steep edges of the Shield, water courses in softer soil were formed, draining to the sea. In this way were born great arterial rivers such as the Mackenzie, the Churchill, the Winnipeg and the Ottawa. There is, in sum, a close correlation between the main fur trade route across Canada and the border of the Shield. Trunk routes sometimes crossed the granite country, but only where a depressed corridor lay, such as to be found connecting the Ottawa with Georgian Bay, or where a great river like the Winnipeg or Churchill spilled from lake to lake as it twisted its way through. In very broad terms, a third of the historic main route to the Athabasca Country lies through granite; a third around the Shield's rim; and a third away from, but parallel to, its rim. Apart from minor detours, no feasible alternative main route offers itself.

Across Canada are three key lakes which serve as hubs or distributing points — one in each of the above great basins: Lakes *Superior, Winnipeg,* and *Athabasca.* Each sits at an altitude of only 600-700 feet above sea level, and has easy communication with the sea. Radiating out from each are six or eight important canoe routes. These three lakes are the keys to the extraordinarily flexible system of canoe travel; they are illustrated in the accompanying sketch maps, and bear detailed study with the map of the main waterways on pp. ii and iii of this volume. The two main gateways are the St. Lawrence and Hudson Bay.

The Precambrian Shield has a further important influence on why the old fur trade routes are still so easily traceable: the routes lay mostly through rocky northern forest, unsuited to settlement or industry. The sixth of Canada which is arable lies almost entirely to the south of the trade routes. Hydro has been the principal agent in obliterating canoe routes. This is fortunately restricted to only a few rivers: the Ottawa, Winnipeg, and Kaministikwia, with the Peace and Columbia in the process of being dammed at time of writing.

Wilderness and climate have combined for hundreds of miles to help preserve the old routes. Areas without roads still use their ancient avenues

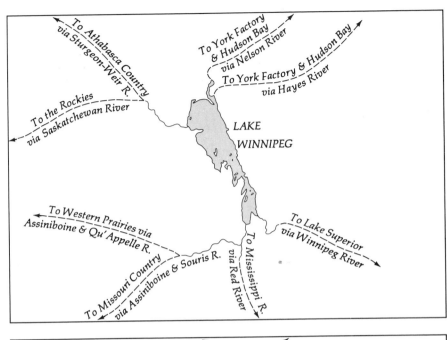

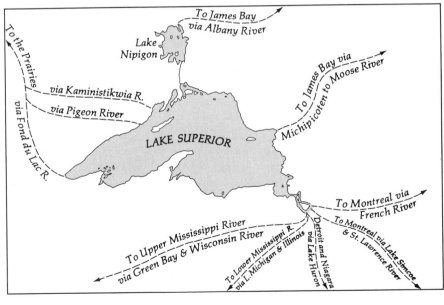

of transportation. The association with granite is an important influence opposing change. Only three centuries have elapsed since the first fur-seekers went over most of these routes, a mere wink in the eye of Time — or in erosion of granite.

Trade routes are traceable from their descriptions in the old journals. The early geographical names are still frequently enough in use to give the general line. The fine points, as to which particular channel or branching the voyageurs used, have to be decided almost entirely by descriptions of portages and rapids. Alexander Mackenzie is helpful in this regard since, besides describing the portages, he recorded their pacings.

The evolution of a portage is an interesting subject for speculation. Speaking personally, I had always assumed that portages were man-made, primitive engineering works, and that game (of which there is always abundant evidence) merely followed man in. After some years of canoeing in uninhabited areas, I am now convinced that game originally made the portages, and that man merely improved them, where necessary. (I am referring here to portages which lie beside a rapid, and not between isolated lakes). There are moose paths along the uninhabited part of Lake

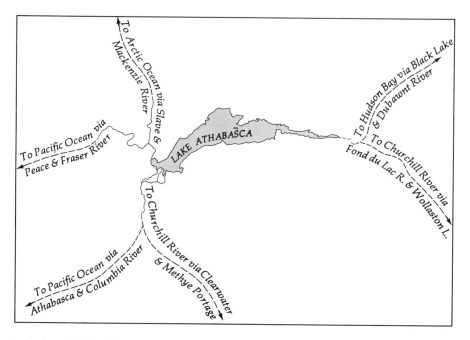

The three "hub" lakes

Superior's north shore west of Michipicoten, exactly resembling in width and appearance the most travelled portages on the Churchill. On more than one occasion near the Arctic Circle, having decided to take a short-cut or a detour by way of a chain of small lakes, we have easily got through. In nearly all such cases, in country inhabited by moose not man, there were good portages, lacking only in ultimate directness. Moose and caribou follow river valleys; where the river is pinched in at a rapid between lakes they will certainly not choose to walk up or down the rapid itself, but along the shore. Like man, they seek a good gradient, going around a steep rise rather than directly over it; like man, big game will go around a bog-hole rather than through it. The three criteria of a portage are: first, directness; second, gradient; third, footing. In only two or three places did the fur companies seriously modify the old Indian portage already present. The packed, weed-free character of a portage path, centuries old, is a combination of the work of animals and man; and the contribution of game may be the greater. Ungulates such as moose and caribou in particular, with their sharp, flat hooves do much for portage maintenance; hence they are a chief agent in portage permanence. In retracing a fur trade route, finding original portages still in their same condition adds a whole new dimension.

The geographical influence may be summed up by saying that the Canadian fur trade was carried on in cold, northern forest areas unsuitable for agriculture, with native populations dependent on hunting. The goods could not have been carried in, nor the furs out, over such distances in an age before modern transportation, were it not for Canada's fantastic system of waterways. In general, the routes were moulded by the Shield, whose character served also to perpetuate them.

PART TWO

Introductory Note

The general navigational, economic, and geographical background having now been brushed in, a better idea may be had of the actual location of the routes.

What is to be ruled a fur trade route? Any river or small stream down which an Indian or *coureur de bois* brought his furs, it might be argued, was a trade route. However, this would plainly violate common sense. Moreover, to write on a subject of that scope in Canada would fill a library shelf. Certain quite arbitrary requirements should first be established. I would list three criteria of a fur-trade route in Canada:

1. One end should be in salt water, or at least tide water — for the great bulk of the trade goods came from overseas.
2. A trade route should include more than merely a single river.
3. A trade route should be more than just briefly transitory or exploratory in character.

Using these yardsticks, there were essentially three fur trade routes in Canada: (a) the Hudson's Bay Company route from York Factory to Fort Edmonton via Norway House; (b) the Montreal fur traders' route from Lachine to Fort Chipewyan on Lake Athabasca; and (c) the various routes attempted across the Rockies to connect the Pacific with either Fort Edmonton or Fort Chipewyan. Routes (a) and (b) crossed each other, with a 200-mile common section. The three routes are treated respectively in Chapters IV, V-VI, and VII.

The treatment of these routes and how they look today will not be uniformly detailed. Where terrain changes have occurred as the result of hydro or industrial development, as on the Ottawa and the Winnipeg, or are immediately threatened, as on the Peace and Columbia, the treatment will be fairly detailed. Where a fur trade route lies in undeveloped country beyond roads, and is therefore still in use, unchanged, the treatment will be more general; examples are the Churchill River, or the Border Lakes route between Lake Superior and Rainy Lake. However, in many cases a route was in use only for 10 or 20 years, or was merely ancillary. Such routes will be discussed sketchily. Non-canoe routes, as for sailboat, horse, or bateau will be touched on as belonging to the ancillary category.

Chapter IV

York Factory To Edmonton

Hudson Bay, as Canada's second main gateway, leads into the interior by such avenues as the Churchill River, the Nelson, the Hayes, and the Albany. By the time that (Sir) George Simpson had taken over, in 1821, the Hayes was clearly established as the mainline. As opposed to the Churchill and Albany, it had easier connections with the interior network of waterways. Its great advantage over the Nelson, emptying into the same estuary, lay in being smaller and thus more suitable for canoes and York boats. The Nelson River, carrying the whole burden of the Saskatchewan, Red, and Winnipeg Rivers, has too much volume of water; moreover, flowing as it does mostly through Precambrian granite, its sharp drops and corners make it dangerously turbulent for small craft. This explains the selection of a minor river, the Hayes, as a trade artery. In fact, in the days when it was used as a fur trade route, it was not even designated as a single river throughout its length: the name, Hayes, applied only to the sixty miles below where the present Gods River flows in. From here upstream to the junction with the Fox River, it used to be the Steel River; and the steep portion between Knee Lake and the Fox junction was known appropriately as the Hill River. Between Knee Lake and Oxford Lake, the Hayes used to be known as the Trout River. Finally, to add to this confusing early catalogue of names, Weepinipanish and Franklin's River were sometimes applied to the portion above Oxford Lake. The Canadian Permanent Committee on Geographical Names has sensibly adopted the single name, Hayes, for the whole river.

At the mouth of the Hayes River, York Factory stands abandoned, a large, bleak wooden building — once the busily throbbing heart of the Hudson's Bay Company's inland trade. Goose hunters on these barren, marshy flats were till recently tearing off its boards one by one to burn for fuel. The National Historic Sites Service has now fortunately taken steps for its preservation, placing guards there during the goose-killing season, fireproofing, and guarding against further structural deterioration.

The 1500-mile trade route from York Factory to Edmonton has so far been changed by man for only one section, above the new Grand Rapids power dam, where the Saskatchewan flows into Lake Winnipeg. The route from York to Edmonton is treated in three divisions: (a) the Hayes River, from its mouth to Painted Stone Portage; (b) the middle section from Painted Stone Portage via Lake Winnipeg to Cumberland House; and (c) the Saskatchewan River above Cumberland House.

To treat the whole 1500 miles from York Factory to Edmonton as a single trade route is simply a geographical convenience, for historically it was not developed as a unit. This volume, however, is concerned more with geographical, than historical, entities.

The Hayes River

The modern voyageur seeking to retrace the old route between Norway House and York Factory should know that the attractive rocky lakeland of the Precambrian Shield, in the whole crescent running from Moose Factory to Churchill, lies well back from Hudson Bay. Another drawback to retracing the Hayes route is a close succession of rapids extending for about fifty miles downstream from Knee Lake — the voyageurs' "Hill River". In this sixth of its total length, the Hayes makes five-sevenths of its total drop to the sea. At the low water of July and August this requires a good deal of wading and portaging. Not having personally paddled this less attractive part of the Hayes below Knee Lake, I remain indebted to others for first-hand information of these 170 miles. A vivid contemporary account such as Sir John Franklin's, on his way (upstream) to his first Arctic exploration in 1819, gives the full flavour of toil and monotony:

> "On the 9th of September, our [York] boat being completed, arrangements were made for our departure as soon as the tide should serve ... The wind and tide failing us at the distance of six miles above the [York] Factory, and the current being too rapid for using oars to advantage, the crew had to commence tracking, or dragging the boat by a line, to which they were harnessed. This operation is extremely laborious ... Our men were obliged to walk along the steep declivity of a high bank, rendered at this season soft and slippery by frequent rains, and their progress was often further impeded by fallen trees."

Franklin, now having gone 120 miles up the Hayes in alluvial country is about to make his first encounter with the Shield. Here the first bedrock is met, causing the first rapid. Between this point and Knee Lake lie most

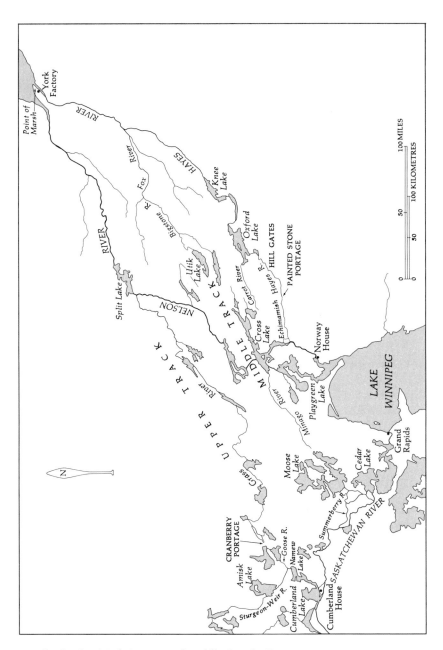

Cumberland to York (upper and middle "tracks")

of the Hayes rapids and portages. Franklin lists twenty, which go by such evocative names as Ground Water Creek, Upper and Lower Burnt Wood, Mossy, White Mud, Swampy, Long, Rocky Launcher, and The Devil's Landing Place. In 1964 all these portage trails were in good shape, though their landings were sometimes screened by undergrowth. What looked to be rollers for boats lay in several places along the portage paths.

Franklin's crew poled, pulled and portaged their way up this long stretch of rapids, while he meticulously surveyed and recorded as he went:

> "We recommenced our voyage, and having pulled nearly a mile, arrived at Borrowick's Fall, where the boat was dragged up with a line, after part of the cargo had been carried over a small portage ... We hauled the boats up several short rapids and carried them over [two portages] ... having proceeded, during the whole day, only one mile and three quarters ... It is not easy for any but an eye-witness to form an adequate idea of the exertions of the Orkney boatmen in the navigation of this river ... Our whole distance this day was one mile and a quarter."

Now entering Swampy Lake, Franklin records that, about ten in the morning, they arrived at the Dramstone, "which is hailed with pleasure by the boats' crews as marking the termination of the laborious ascent of Hill River. We complied with the custom from whence it derives its name". There was still, however, an eight-mile stretch between Swampy and Knee Lakes, involving four portages. Knee Lake on the Hayes gets its name, as other Knee Lakes do, from a sharp "knee" in its middle, where the lake changes direction by nearly 90°. Granite-girt, island-studded, and ample, this lake excited comment by its upstream travellers emerging from a fortnight of gloomy toil. Few failed to mention Magnetic Island, half a mile from the "knee", whose ore-body renders compasses useless, when close by.

The topmost section of the Hayes, above Oxford Lake, is for the most part an attractive alternation of lake and stream. This stretch contains two locations of chief interest, Hill Gates and Robinson (or White Fall) Portage. Hill Gates is a gorge three quarters of a mile long, so narrow in spots that there was not room to row the boats. Robinson Portage, a mile long, was (with Grand Rapids Portage) the longest portage between York Factory and Edmonton. In 1956 we found relics of a railway running on iron-plated wooden rails. At its upper end lay submerged a rusty, flat four-wheeled vehicle which would take a York Boat or its load.

Painted Stone to Cumberland House

Why the rowers of York Boats, when there is a direct and continuous flow of water from Lake Winnipeg to York Factory by way of the Nelson, preferred to lug their boats over a divide, and take to the Hayes River instead is partly explainable by the bad character of the Nelson, but also by the facility of movement over the divide. This is offered by a curious little river, the Echimamish (the-river-that-flows-both-ways). The Echimamish, with barely detectable current for about forty miles, links the Hayes River with the Nelson. In totally flat, swampy country the Echimamish boasts the unusual feature of rising, not at one end, but in its "middle"; two streams, from north and south, meet in a beaver pond which flows out both west and east — respectively into the Nelson and the Hayes.

At Painted Stone Portage, canoes and their cargo are carried for twenty paces, *not* over a divide — as is commonly stated in the early journals — but from the river (flowing eastward now) over to the head of a long, narrow bay of the lake. This short portage avoided a shallow, rocky rapid in the stream.

At the first beaver dam on the Echimamish the canoes are simply dragged *up* over the dam; and a long beaver-flooded section follows, still proceeding westward. Shortly after the two source streams enter, another beaver dam is encountered, where the water now is flowing *down*. Two lower beaver dams are met on the way toward the Nelson. The paddler in a dry summer would wish there were more, for below each dam there is sometimes hardly enough depth of water to float the canoe. The beaver here serves as an aid to navigation. When beaver became exterminated in this area, the Company had to go into the dam-building business itself, of which traces remain today. Fortunately, whether heading up or down, the brigades generally missed the low water of midsummer.

The Echimamish, as a result of flowing through peat bogs, is very dark in colour. Between Hairy Lake and its mouth in the Nelson it is called Blackwater Creek, from its contrast with the Nelson's tawny flood. On the Nelson, no rapids are met until Sea River Falls; this was one of the Nelson's early names. The Nelson continues to have strong current until it broadens out in Playgreen Lake, on which sits Norway House.

The present Norway House was built in 1826. (The first House, built opposite Warren Landing at the outlet of Lake Winnipeg, burned down).

A photograph entitled "Hauling a York Boat over the Robinson Portage, 1878". Five men are hauling, like draught animals, on the line; another ten men are lifting at the gunwales. The photograph illustrates the awkwardness of the York Boat in portaging.

This is one of the most attractively situated Hudson's Bay posts in the north, screened from Little Playgreen Lake by a rocky shoulder. Its clean white buildings are set off with groves of jack pine. Norway House became a key in the Company's transportation system. It was also the administrative centre of the Northern Department, a meeting place for the Northern Council, and a major depot for collecting furs and goods for onward shipment.

The route from Norway House to Grand Rapid at the mouth of the Saskatchewan River lay along treacherous Lake Winnipeg's north shore. When the wind was fair, the York Boats sailed — sometimes night and day. For forty miles westward along the lake coast there were no landing places in a strong south wind, for then the surf beat against the foot of the high clay cliffs. With a north wind driving the shallow waters southward, a narrow sandy beach was exposed, affording a landing place for boats. In Lake Winnipeg's north-west corner lies Limestone Bay, separated from the lake by a low sandy spine fifteen miles long. In a bad south wind, the boats could be carried across its narrow isthmus to proceed in its lee. Limestone Bay brought the boats to within forty miles of shelter, in the mouth of the Saskatchewan River. The section of the Hudson's Bay Company Edmonton route up the Saskatchewan River between Grand Rapids and Cumberland, a section shared in deadly rivalry with the Northwesters, is described in the next chapter.

Hayes and Echimamish Reading

| **12** | I 41-66 | **19** | 99 fn., 246-51 | **57** | 93-97 |
| **14** | 147-52, 189 | **43** | 272-4 | | |

From Cumberland House to Edmonton

The location of Cumberland House was a shrewd selection by Samuel Hearne as the Hudson's Bay Company's first major inland post in 1774. Via the Saskatchewan it had connections with both Lake Winnipeg and the Rockies; here too the Northwesters turned north off the Saskatchewan, heading for the Athabasca. Thirty miles north, another route branched off to the lower Nelson. Cumberland House was a crossroads. It sits on an island in Cumberland Lake, a broadening of the Saskatchewan River.

The Saskatchewan in 600 miles between Cumberland and Edmonton has no rapids which could not either be lined up or run down. The main menace to navigation is its shifting gravel bars in muddy water. There is, however, one particularly strong 125-mile reach of current extending upstream from 25 miles below Nipawin to near Prince Albert, where the Saskatchewan drops steeply off the middle prairie level to the lowest (Manitoba) level. Duncan M'Gillivray, ascending the Saskatchewan in 1794, wrote of this section:

> "Towards night arrived at the beginning of the tracking ground where we camped. This part of the River is an object of terror to the whole band, the usual mode of navigation is here rendered useless by the Strength of the Current, which makes it necessary to haul the Canoes up along Shore with a line ... for the space of six days on an exceeding bad Road."

Shortly above Prince Albert a welcome change occurred in the diet of the upgoing voyageurs, when the brigades reached *La Montée*. The country here changes character as the open prairie is approached, and the "gentlemen" *mounted* (hence the name) to kill buffalo. Keeping generally abreast of the brigade, the hunters were able to keep the crews provisioned with fresh meat. Duncan M'Gillivray hailed the change: "What increases the beauty of this agreeable prospect is the inumerable herds of Buffaloes and various other animals that inhabit those delightful regions of plenty, which always afford a seasonable supply of food to our men, who are now almost worn out after the fatigues of a long and toilsome voyage."

There is little to attract the modern voyageur to re-paddle the murky Saskatchewan. The absence of portages reduces awareness of the past. Though the scenery has changed not at all from the days of the fur trade, high banks along much of the river restrict ·the view; and the scenery tends, as on any long river, to become monotonous. Herds of buffalo and other big game no longer enliven the scene.

There remains in the present chapter but to sketch in a few ancillary routes and connections — including some only temporary ones. Throughout the mainline route from York to Edmonton, York Boats were used by the Hudson's Bay Co. from 1797. The North West Co. and other Montreal traders used North Canoes. Both canoes and York Boats plied all the bigger Prairie rivers west of Lake Winnipeg. In the smaller and rockier rivers north of the Saskatchewan, the unportageability of the York Boat discouraged its use. After the North West Co. had been absorbed, the Hudson's Bay Co. had the advantage of a pool of experienced canoemen to draw from, and was able to use either York Boats or North Canoes, as terrain or economy dictated.

43

Using Cumberland House as a pivotal point, water connections in all four directions should be noted. As fur-trade routes, the Prairie rivers to the south-west need sparse comment: this was not canoe, but horse and Red River Cart, country. The main contribution of the Prairies to the fur trade was not furs, but pemmican and buffalo hides. Moreover, as outlined earlier, the independence which the buffalo economy gave the Prairie Indians made this not always healthy country for fur traders to travel in, at least not very far from their forts. The brigades from Edmonton, George Simpson recommended in 1824, should be "Seven Boats and Forty Five Men . . . which the Natives must respect". The North Saskatchewan formed roughly the boundary of the "strong woods" region where the furs were harvested. The Souris, Assiniboine, Qu'appelle, Swan, Red Deer, and Carrot Rivers all bore fur traders' canoes, but they were ancillary routes. The traffic on these rivers consisted more commonly of roughly-built boats, sometimes made for only a one-way trip, bearing tons of pemmican to the "refuelling" posts.

To the south of Cumberland House, and flanking Lake Winnipeg, lay a route through Lakes Winnipegosis and Manitoba, which would deserve less attention in this book, had the Canoe Pageant in 1967 not been routed through this way. Fur trade along this route was strictly local. Through traffic of freight-canoes and York Boats would be deterred on this route by three long portages: *Mossy Portage* from Cedar Lake on the Saskatchewan to Lake Winnipegosis, four miles; *Meadow Portage* across to Lake Manitoba, a messy mile and a half across an isthmus to avoid the Dauphin River's long sharp elbow and shallow rapids between these two lakes; and *Portage la Prairie*, five miles at a minimum, even making use of small streams, in order to cross from Lake Manitoba to the Assiniboine River. The Hudson's Bay Co. usually avoided Portage la Prairie by entering Lake Manitoba from Lake Winnipeg, via the Dauphin ("Little Saskatchewan") River and Lake St. Martin. From only Norway House was the Mossy Portage route to Swan River used. The Canoe Pageant in 1967 chose to travel via Lake Manitoba instead of the regular route via Lake Winnipeg simply because, as a Centennial Pageant, it had to pass through the cities of Portage La Prairie and Winnipeg.

The "Upper and Middle Tracks", Cumberland to York

The *main* canoe route between Cumberland House and York Factory via Lake Winnipeg, described above, did not come regularly into use until

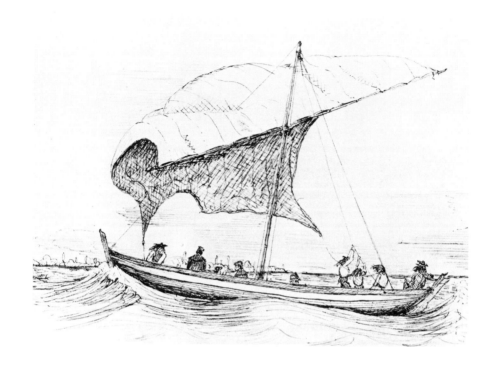

York Boat on Lake Winnipeg, 1848, from a drawing by George E. Finlay.
A fur canoe was rigged similarly.

after Cumberland House was established. In the century between 1670 and 1774 the Indian middlemen had used another route. The maze of rivers which lay between Cumberland and York offered considerable choice of routes, but two in particular were most commonly used as being shortest and easiest. These came to be known as the Upper and Middle Tracks (or Roads), and are illustrated in the accompanying sketch map (page 38).

The "Upper Track", from Cumberland, started up the Sturgeon-weir, branching north-east a few miles past Namew Lake, up the Goose River and over Cranberry Portage where the Grass River is entered. Grass River leads down to the Nelson, joining it just above Split Lake, from which point it is only about 200 miles to York Factory. Cranberry Portage, forty miles south-east of Flin Flon, provides an easy modern approach by road or rail to the Grass River, which is still unchanged and attractive canoe country. There would be little point in going below Split Lake on the Nelson, even without the destruction to be caused by the new dam: going up or down, the Nelson is not a canoeing river.

The "Middle Track" was the canoe route principally used by Indian middlemen coming and going between York Factory and the Saskatchewan River, prior to the establishment of Cumberland House. From below The Pas on the Saskatchewan, this route passed down the Summerberry River and up into Moose Lake; crossing a low divide it went down the Minago River to Cross Lake on the Nelson. From the east end of Cross Lake the "Middle Track" branched, one route passing north-east directly to Utik (Deer) Lake and down the Bigstone and Fox Rivers to the Hayes. Henday and Cocking used this route on their way west. The other branch passed east from Cross Lake through Walker Lake and Carrot River, joining the Hayes in Oxford Lake. The Middle and Upper Tracks were abandoned as trade routes with the building of Cumberland House. Henceforth the furs and trade goods would be carried in freight canoes, and (after 1797) in York Boats, which required much bigger waters. The Middle and Upper Tracks were suitable only for the small canoes of the Indians.

These two earlier routes lie wild and unchanged today. However, any-one seeking to retrace them is warned that the two divides, (a) between Moose Lake and the Minago River, and (b) between Cross and Utik Lakes are now practically impassable and grown over. The Indians in earlier years may possibly have been assisted by more beaver dams on the creek-lets; or more probably the Indians with their light canoes were prepared to walk farther, and their regular traffic kept a way open. The Grand Rapids

Dam has now raised the level of Moose Lake seven feet.

However, the full sense of early travel in this area can be enjoyed on a circle route which I took in 1956; this embraces the best of the Hayes and the Middle Track, and avoids the worst of both. The route starts at historic Norway House, reached by regular scheduled flight from Winnipeg. Descending the Nelson to the Echimamish, then passing down the Hayes to a little below Robinson Portage, a delightfully sequestered chain of small lakes leads to Gods Lake. Two-mile Bayly portage through muskeg shakes off the motor launches of Gods Lake and leads over to the Wolf River and on into Knee Lake, thus hitting the Hayes just above its "Hill River". Upstream then on the Hayes to Oxford Lake brings one into direct communication with Cross Lake by way of the Carrot River section of the old Middle Track. The Carrot River, flowing through small rocky lakes dotted with islands, has a rare charm.

The modern voyageur travels in this area with Kelsey, Henday, Franklin, Cocking, Thompson, Fidler, Simpson — to mention but a few. Kelsey, the first white man to view the Canadian Prairies, almost certainly would have been guided into the interior by the Indians, in 1690, along what would be their principal route for bringing down furs. Because Kelsey chose to describe his route in atrocious jingle, however, it has provided a field day for historical and geographical conjecture. Kelsey's "Deering's Point" is generally acknowledged by scholars to be somewhere on the Saskatchewan River, and a very likely location would be The Pas. The route by which he arrived at The Pas is cloaked in the obscurity of Kelsey's celebrated doggerel:

> "Distance from hence by judgment at ye lest
> From ye House [Y.F.] 600 miles southwest
> Through rivers which run strong with falls
> Thirty-three carriages, five lakes in all."

The south branch of the Middle Road gives exactly five lakes (Knee, Oxford, Walker, Cross, and Moose), and the requisite number of portages. The Pas would broadly fit Kelsey's mileage and direction from York Factory. The Pas also offers easy access to the Assiniboine Indians, a point Kelsey stresses. All this, supporting the strong presumption of Kelsey's having accompanied Indians on their known regular trading route to and from York Factory, makes the Middle Track, south branch, look very plausible as Kelsey's route to the Prairies.

Chapter V

Montrealers' Mainline:
1 / Lachine to Grand Portage

The existence of this "Trans-Canada Highway No. 1" is known to most Canadians. For years it was Canada's economic spinal column — and even after 1821 when the western freight traffic shifted to York Factory (and later, St. Paul, Minn.), it remained the only route for mail and passengers to cross Canada till the railway came in the 1880's. Its eastern half is parallelled by modern highways, so that people are able in many places today from their car window to share with the early voyageurs the scenes they viewed as they paddled or portaged.

The Ottawa River

Lachine, rather than Montreal, was the start of the line, since horses were cheaper than men for getting heavy goods past the Lachine rapids. The big *canots de maître* started off about the first of May, when the Ottawa had become quite free of ice on its bigger lakes. The start was from the head of the rapids. The building of the first Lachine canal by 1824 obliterated the original canoe landing place. However, there is enough circumstantial evidence to suggest where it must have been. The same conditions of current which dictated where to begin the first canal also determined where canoes should be loaded. A short distance below the head of the canal, and opposite a large convent, is an old building known to have been a Hudson's Bay Co. warehouse. The canoe launching area was probably between this warehouse (still occupied, but converted to living quarters) and the head of the old canal.

At Ste. Annes, sixteen miles farther, the canoes were paddled *demi-chargé* up the mild rapids. The stone convent at Ste. Annes, built near the head of the present canal, was recently examined by an architect, who found considerable evidence to support local rumour that its bottom storey was in fact the little church where the voyageurs landed and left their mite, to

W. H. Bartlett's print allegedly depicting "Working a Canoe up a Rapid" (from *Canadian Scenery*). The voyageurs' hats and the canoe are somewhat inaccurate, but the scenery so precisely portrays a scene the author knew well from camping and picnicking that he is convinced it was the foot of the (now drowned) rapids at Portage Dufort on the Ottawa, with the portage starting at the right. The canoe has come in to the landing and is being unloaded in the water.

receive the priest's blessing for their hazardous journey to the *pays d'en haut*. Lake of Two Mountains now looms; and the voyageurs usually timed their departure from Lachine to be able to make their first camp near its upper end, where each canoe crew received a keg of rum (or else "high wine"). It was always a rough day, following that "morning after."

The canoes after paddling Lake of Two Mountains came to the twelve miles of rapids, in three sets, comprising the Long Sault. Old maps support the knowledge that the canoe route held to the north (Quebec) side. The voyageurs got past these rapids mostly by tracking, but occasionally finding in quieter sections a place to paddle *demi-chargé*. Alexander Mackenzie indicated that "in this distance are three carrying-places [the "*Trois Ecores*"], the length of which depends in a great measure upon the state of the water, whether higher or lower". It was at the foot of the Long Sault, as he was on the point of portaging, that Adam Dollard des Ormeaux made his celebrated stand.

The sixty miles of the Ottawa from Hawkesbury to Ottawa have been changed merely from fast water to dead water, the headpond of the Carillon Dam having raised the water-level as far back as the foot of the Chaudière Falls. There is nothing of note in this distance, except that above the Long Sault the voyageurs first came in view of the edge of the Precambrian Shield, which would lie to their north for most of the way to Grand Portage. To those of Canadian birth these were friendly hills reminding them of home; to visitors from Europe, whether French or English, the rocky Laurentian hills looked sombre, barren, and forbidding.

Now came a strenuous seven-or-eight-mile stretch of rapids and fast water between present Hull and Aylmer, P.Q. The rapids were the main Chaudière, the Second (or Little) Chaudière, and the Deschênes. All were passed on the north shore.

There were two different lower landings at the Chaudière, causing a Y-shaped portage. Going upstream in early May with the Ottawa in spate, approach to the foot of the falls was quite impossible. A long, narrow, steep-sided gorge in the limestone, however, branched off just before this; its flat water allowed the canoes access almost to the falls, where a steep path led to the top, requiring every man in the crew to help haul up the canoe. Unfortunately this little gorge is being fast filled up with waste and garbage, and its water has been diverted. From its head, the portage path began, and probably lay along the line of the present railway, thence across Brewery Creek to a point beside the railway tracks, opposite the

This photograph of one of the type of canoes used in the Centennial Canoe Pageant was taken in 1967. It is a simulated North Canoe, and is shown running the Deschênes Rapids on the Ottawa as the voyageurs would run them, except that the canoe is not loaded.

rise to higher ground. The portage lay so close to the falls that, with an on-shore wind, the voyageurs were drenched in spray. On the way back in late September, the Ottawa would be at its lowest. A shorter portage was thus possible, branching off to a small bay at the very foot of the fall, and close below the present Chaudière bridge. This was known as the "summer landing"; at the head of the little gorge was the "spring landing".

A delight to those retracing the old route today is in store at the Second Chaudière Portage, a mile upstream, for here slumbers unchanged one of only three short stretches of original portage left on the whole of the Ottawa River route. It includes two or three sets of crude stone steps and a low causeway, all built by the early voyageurs. This old portage, half a mile long, lies partly in and partly below Brébeuf Park in Val Tetreau, within view of the Champlain bridge. From the head of the portage the voyageurs paddled across a small bay to a beach to get ready for lining and poling four miles to the Deschênes Rapids. These rapids were portaged in a straight line from a small bay at their foot to a bay just above. Coming downstream, the Deschênes rapids would be run.

A ridge of granite lies across the Ottawa at Chats Falls, once causing (before the power dam) a line of beautiful waterfalls, a mile wide. Another unspoiled section of old portage crosses the *second* island from the north shore. As numbering is confused by a partial canal-cutting, it should be added that the island is identifiable by an elbow in the dam above, and by three small islands off the foot of the portage. Bartlett depicted this portage in a familiar print ("Portage des Chats"); and the portage is also recognizable in another of his prints ("Burial Place of the Voyageurs"). The dam and its rockfill now obscure the upper portion of the portage.

Before the dam was built, there was above the falls a mile or two of very fast water through islands. Some hair-raising accounts tell of the brigades going up this section with the Ottawa in flood, the voyageurs often pulling the canoe up by means of the branches along shore, with the falls roaring below them.

At the upper end of Lac des Chats the course of the river is bridged by a small archipelago, between whose islands the current flows fast. This is the Chenaux. Local legend hints that a portage was required here, but it is not substantiated in the journals. The channel closest to the Quebec side was probably the one travelled; and, depending on water level, poling or lining might have been necessary at a couple of points. At high water, Père Belcourt spoke of it as "un courrent rapide qu'on ne monte qu'en se tenant aux branches".

52

In this context should be mentioned the route the Indians used, and guided Champlain over, to bypass the bad section of the Ottawa between Lac des Chats and Lac des Allumettes. Unencumbered by freight, and travelling in small canoes, the Indians left the Ottawa on the Ontario side just above the Chenaux, climbed the hill and passed by way of a chain of twelve small lakes, including Olmstead Lake, over to Muskrat River which drains out through modern Pembroke. Along the Ottawa Valley is frequently encountered the inaccurate term, "Champlain Trail", as applied to Canada's main fur-trade route — implying that Champlain either discovered it or was its only traveller. Only the short, little-lakes route described above could accurately be called "Champlain Trail" — as also could his route from Huronia by the Trent Canal route to attack the Iroquois in 1615. There seems no justification for singling out one name to commemorate a generally used turnpike extending 3000 miles, especially when, in Champlain's case, he had travelled only a sixth of its distance.

The *voyageurs* in the freight canoes, above the Chenaux kept to the Ottawa, even following it around a big bend involving twice the paddling distance. The reason was that freight canoes could move at four miles an hour even against a current, whereas on portages their net rate worked out at no more than half a mile an hour. The Muskrat River route like the Middle Track in Manitoba was strictly for lightly-laden Indian canoes, not *canots de maître*.

Within sound practically of the Chenaux are the rapids at Portage Dufort — "où périt", according to Père Belcourt's journal, "un voyageur de ce nom, lequel, par extravagance ayant tenté de le sauter, paya de sa vie son imprudence" — an expensive immortality. This explanation of the name casts doubt on the local legend that the "Du Fort" alluded to Fort Coulonge, thirty miles upstream. I knew this beautiful rapid on the Ottawa in the 1930's before it was drowned. It cascaded down through many channels among big trees. Bartlett's print "Working a Canoe up a Rapid" (falsely so entitled) very plausibly is its lower landing, showing voyageurs unloading the canoe and starting off along the portage path to the right. Bartlett had passed this way in the 1830's; and the scene matches.

The Portage Dufort dam also flooded out rapids which caused the *décharges* du Sable and D'Argis, and the Mountain Chute. The Ottawa River splits here, rounding Calumet Island. The voyageurs chose the eastern channel for less frequent carrying, most of the drop being picked up in the single Grand Calumet Portage. This was the longest portage east of Grand Portage, a mile and a third long. The portage path was still faintly traceable in 1967, starting well below and almost around the corner from the foot of the dam, on the west shore. Ascending gradually through mostly cedar from a little cove whose mouth is now choked off by rubble, the path does an easy traverse of the hill, to pass above two steep ravines. It is soon interrupted by the western bay of the headpond. The path is picked up again across this bay in an old abandoned roadway in open ground; slowly climbing, this eventually merges with the main road, which lies on the old portage path. From the crest of the main road, the portage then followed the course of the *old* road forking off to the right. The upper landing of the portage would probably have been where the road comes closest to the river, at a sharp corner and foot of a hill.

The next problem in relocating the old route related to the Allumette Rapids. Which of four channels was used? The channel taken by the voyageurs going upstream was the small one south of Cotnam Island, against the main Ontario shore, where Dr. Bigsby had described "pushing up another passage like a sewer or tunnel". Though some rock has been blasted at one spot, the short portage is otherwise unchanged and in use. Coming down river in September, there is little doubt (though I have failed to find actual record) that the brigades shot the Allumette Rapids in their main (north) channel, the "Timber Snye", where a safe canoe course passes close to the north of the little island in the rapids.

At the upper end of Lac des Alumettes the voyageurs came for the first time on their journey into close contact with the granite of the Precambrian Shield, rising straight from the water. High pine-clad rock rising to 500 feet above unfathomable depths gave this section the name *"Rivière Creuse"*, perpetuated today in the town of Deep River. There was now a feel of getting really away from settlement and into the *pays d'en haut*. Nearing this new narrowing, this change in the river's character, was a commodious campsite, a long sandy point among pines. Here was the traditional occasion for a voyageur ceremony, "baptising" any novices in the crew, who were then required to "stand" a *régal* all round. *Pointe au Baptême* remains clearly identifiable and unchanged, just below the Chalk River Plant of Atomic Energy of Canada, on the Ontario side.

The next historic portage was at Des Joachims (Ottawa Valley pronunciation: day-swish-a), where a power dam now backs up the Ottawa for 70 miles. The river here does a big S-turn, where two miles of thundering rapids once dropped. The old portage made use of two bays

54

and a little lake, to cut down the total portaging distance to less than a mile, in two sections. To locate the old portage it is necessary to start from the little L-shaped bay immediately north of the village of Rapide-des-Joachims, at the foot of the big island lying between the two channels of the Ottawa. The first part of the portage is untraceable owing to the houses, but then an old trail can be picked up, heading for the south-east corner of a little lake, now called Bell Lake, and maintaining a consistent gradient. Gullible novices, always the butt of voyageurs' jokes, were told as they crossed this tiny lake of a voyageur who was once pinned down by the wind here for several days. The voyageurs then tackled the second Portage des Joachims, whose approach according to Garry started "at the end of a small channel, the sides of the canoe touching the banks". This little channel, still evident, leads from Bell Lake to the present dirt road passing near the lake. Branching off this road just past Bell Lake is what looks to be an old and partly overgrown lumber road, which from its direction, gradient and length would seem to coincide with the old portage path.

Beneath the waters of the Des Joachims headpond lie what were once some of the Ottawa's most attractive and turbulent rapids, Roche Capitaine, Deux Rivières, the Trou, and l'Eveiller, commemorated now only in a village name and an Ontario plaque beside an overlook on the highway.

The Mattawa River

The Mattawa is a little gem of history: rocky and rugged for nearly all of its forty miles, it has defied cultivation or settlement, and remains for most of its length unchanged in every way. Yet within earshot on a still night the throb of diesel engines and the roar of heavy transport trucks come through from the railway and highway flanking the river to the south. This river, early in the journey westward, caused much comment by the old diarists.

I have seen no record of portaging the small Mattawa Rapid (at the junction with the Ottawa); Nicholas Garry said that it was lined up. Now it has become victim to the Des Joachims Dam. The next two portages on the Mattawa have likewise gone — under the headpond of a small dam. The first of these is *Plain Champ*, (which also appears as *Plein Chant, Plain Chant,* and *Plein Champ*). I prefer the "champ" over "chant" because of common sense; this is also supported by the two most reliable chroniclers, Mackenzie and Macdonnell. A "plain champ" was just a flat field, a feature which would certainly attract notice in this area. The second casualty of the Plain Champ Dam is the *Décharge des Roses*. The map-makers have unfortunately confused its location with the mouth of the Amable du Fond, a tributary stream, which would leave it out of proper sequence in the early accounts.

At the junction with the Amable du Fond is, rather, the *Décharge Campion* (which has come out "*Champagne*" and "*Compagnie*" in one or two journals); confusion over names is understandable in the case of a passenger trying to catch a name merely verbally, and sometimes over the roar of the rapid. The "Campion" for whom the *décharge* was named is very likely to have been Etienne Campion, Alexander Henry's assistant in the 1760's. Then in succession were the Portage des Roches, the Portage des Paresseux, the Portage de la Prairie, Portage de la Cave, the Portage Talon, Portage Pin de Musique, Portage de la Mauvaise Musique, and Turtle Portage — *eleven* in all. The Ontario Department of Lands and Forests in 1967 did an imaginative and commendable job in brushing out and marking the portages on the Mattawa. It is unfortunate only that the introductory statement on each of the signs reads: "One of *fourteen* historic carrying places along the Mattawa . . .". Also the dates, "1610-1835", mar the signs' historical accuracy, for the use of these portages not only goes back thousands of years, but continued until a century ago.

An interesting location is the *Rapide*, or *Anse, des Perches* on approaching Pimisi Lake just below Talon Falls. It derived its name from being the spot where the voyageurs, with some ceremony and much joy, threw away their setting poles *(perches)*. This signalled the end of upstream rapids for the rest of their journey. Two important inferences are possible: (a) this rapid was poled up, not portaged; (b) the course ahead clearly lay through Robichaud Lake, sitting between the two Musique portages, and not via the rocky Mattawa River above Talon Lake as some have conjectured.

The last "portage" on the Mattawa, Turtle Portage, requires explanation for anyone who should fail to notice a small dam at the foot of Turtle Lake, just below. Before the water was thus raised, Belcourt speaks of this spot, in passing on May 9 (i.e. at very high water) as offering "passage de la largeur du canot seulement" (i.e. six feet). Macdonnell, passing at lower water six weeks later, had to portage — confirmed also by Mackenzie and others. Today flooded out, it is just a tight, rocky gut.

Trout Lake is the head of the Mattawa. In order to get to Lake Nipissing, a divide now must be crossed, over a series of granite ridges, sandwiching bogs. The accompanying sketch map illustrates how this was accomplished. First there was a 1500-yard portage, flat, but winding, over the low height of land. This portage led out of the pond across the highway south of Dugas Bay on Trout Lake. An Ontario historical plaque marks the point at which the canoe route crossed the Trans-Canada Highway. The Height of Land Portage, separating Ottawa from Lake Huron waters, set the big canoes in a beaver pond on the tiny tributary of the La Vase (i.e. muddy). Were it not for beaver damming this trifling stream, the portage would have been considerably longer — which apparently was actually the case following extermination of local beaver. Except for two short portages, a succession of pond and creek now carried the big canoes — sometimes with branches brushing their sides — down to Lake Nipissing. It is still possible to paddle over this section of the old route, for the beaver are back, helpfully damming the same old spots, and converting an unnavigable creek into a canal. The whole distance from Trout Lake to Lake Nipissing is about seven miles, of which five are still paddleable.

Lake Nipissing, being shallow, becomes dangerously choppy in a high wind; but the course lies not far from its south shore, and there are many islands. Though local knowledge seems to be blind to its presence, the old Portage of the Chaudière des François was still traceable and in good shape in the nineteen-fifties, starting from a flat rock in a bay to the south of a freshly-blasted outlet and leading to an obvious long bay in still water below the outlet. Here the French River begins.

French River

The French River, seventy miles long, was just a pleasant day's run for the voyageurs, going downstream following its south channel. At the high water level which they had for their downward passage around the end of May, there were but two portages, Recollet Falls, and the Petite Faucille close to Lake Huron. At Recollet Falls, the slide constructed about 1955 for the passage of motor boats lies on the old portage, fifty paces long. If he hopes to recreate the atmosphere of early fur-trade days on the French River, the modern voyageur should avoid this scenic river in the vacation months of July and August, when much of it is churned by

Frances Hopkins's "Running a Rapid on the Mattawa River" shows a lightly-laden express canoe carrying her and her husband, who was Secretary to Sir George Simpson. Note the *avant* and *gouvernail* both standing. The *milieux* appear to be braking with their narrow paddles.

The North Bay divide

speedcraft and polluted by an oily scum; cottages line the shores of the French principally in the quarter of its total distance lying between the two railroad crossings, while just above the C.P.R. high bridge several resorts are clustered. June or September are the months to paddle this river; and a delightful trip can be planned by leaving one's car with an outfitter just below the C.P.R. bridge and having him truck the canoe party to Wolseley Bay. Paddling down from there to Georgian Bay, all the best of the French's scenery and rapids can be enjoyed. A great advantage is that, instead of then fighting current back up the French to one's car, the Pickerel River, flat, and parallel to the French, can be ascended to a couple of bays above the upper railway crossing, where flows an overspill from the French. Here a single, steep 400-yard portage puts the party back on the French River, about three miles above the car and starting point of the trip.

The French River, counting the outlet that it shares with the Pickerel, has four mouths. The voyageurs were interested in travelling by the route which would protect them longest from the open waters of Lake Huron, which of course meant the westernmost outlet. The search for Voyageurs' Channel on the French, however, is complicated again by the map-makers, who for a period labelled the wrong channel, and more recently show the correct channel as being blocked off and impassable. The keys to identifying the correct Voyageurs' Channel are two: a gut, and a curving rapid causing a portage of only 25 yards. The terrain in the area affords many locations where narrow channels race between smooth gneissic rock. Contemporary accounts describe the particular one as: "three to five miles from Lake Huron"; "not more than 10 or 12 feet wide and 100 yards long"; "scarcely to allow the canoe to pass"; and "our canoes flashed through it almost in a moment". This gut was known as *La Dalle* (the Eave Trough). The curved rapid was known as the "Petite Faucille" (the Little Sickle), where the lift-over Mackenzie said, was a mere twenty-five yards. There is, in fact, one channel which exactly includes these two features. It is the most direct western outlet of the French, leading to a long protected course westward then southward, into Georgian Bay. The channel is unmistakably identifiable in a map of the 1850's, and labelled there as "the principally travelled route" of the time. No other channel that I have found in the area fits the description of these two key points, nor matches the 100-year-old map. This channel in 1967 was still navigable at extremely low water in September, the month when the voyageurs would pass, homeward bound. Muskets, shot, and other fur-trade hardware were found in 1971 by amateur divers immediately below the Petite Faucille Rapid, thus further authenticating this channel. The only sure way to find the correct channel is by coming down Western Outlet and making for the lowest land in view, just to the right of the motor-boat channel. The Voyageurs Channel starts in an obscure rocky bay and heads southwest. Almost immediately the canoe is swept into *La Dalle*. Straight ahead and soon, the sickle-shaped *Petite Faucille* blocks the way, and a short lift-over leads to a continuation of the channel. This then comes out on a wider strait, where the voyageurs commenced their westward course, crossing the Old Fort Channel at right angles and taking advantage of maximum island protection out to Georgian Bay. (See accompanying sketch map)

Another place of interest lay where the voyageurs set out for the lake, a grassy meadow *La Prairie des Français* which, said Alexander Henry, "varies for a short space the rocky surface, which so generally prevails". Because of the almost total absence of such spots on the rocky French, commodious enough for a whole brigade to encamp, this was a favourite camping ground where, too, canoes were repaired. Such an area still exists there.

Lachine to Georgian Bay Reading

3	I 129-43, 150-8, 165-71	15	71-85	27	36-43
14	92-107	20	15-32	54	(first instalment)

The North Channel of Lake Huron

The 200 miles from Sault Ste. Marie along Lake Huron's North Channel to the mouth of the French River — and more especially the portion of it east of where the highway comes down at Blind River — is extraordinarily beautiful canoeing: hundreds of rocky islets; for miles a 500-foot range of quartz-blanched hills following the coast; and crystal-clear water. It is, of course, easy of access by road at a dozen points. The prevailing wind is westerly, which would suggest paddling it in reverse progression to that described here.

The voyageurs on coming out of the French River hugged the mainland, being able for the most part to take advantage of a screen of small islands, often a mile or two deep. If there were a bad wind they used the screen; otherwise it was faster to stay just outside. Almost immediately from the

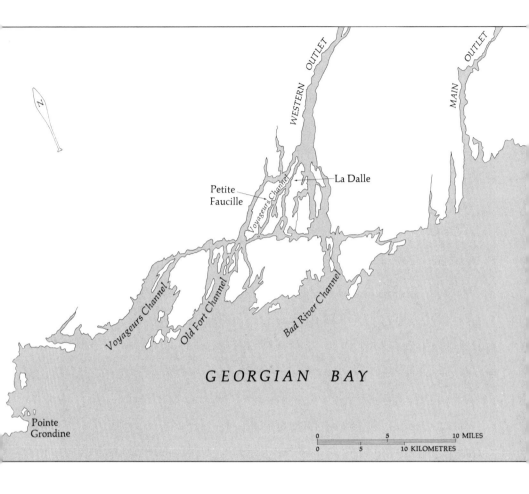

The Voyageurs Channel, mouth of French River

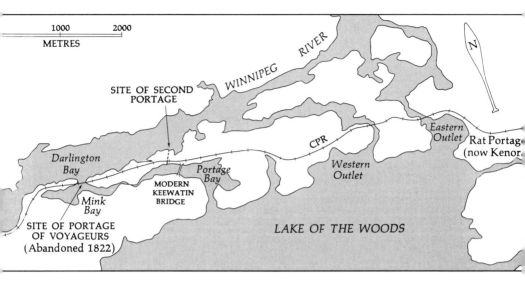

Rat Portage and Outlets of Lake of the Woods

mouth of the French they had to round an exposed point, the *Grondine* (moaning) which got its name from the ceaselessly surging swell among the big rocks.

Shortly after Killarney, by a big traverse the canoes headed for a narrow strait between islands, just wide enough at its head to take a *canot de maître*. Near the upper end of this channel lay an historic curiosity, *La Cloche,* and everyone talked about it. The name "Cloche" locally dominates the map: there are La Cloche Mountains, Peninsula, and Channel; there are South La Cloche Mountains, Great La Cloche Island, and Little La Cloche Island — to say nothing of the La Cloche Indian Reserve. All this on account of a *cloche,* or bell, according to Dr. Bigsby, "so called from some of its rocks ringing like a bell on being struck". He said the name "particularly applied to one loose basaltic mass lying on the shore". It was an object of wonderment to travellers and voyageurs, who had to pass within a few paces of it. I had personally been somewhat sceptical of a rock ringing like a bell, and when paddling through was interested to try to track down this curiosity. We were fortunate in finding with no trouble a man who had seen it and knew its location. He took us to a mass of dark basalt, obviously an "erratic" deposited by a passing glacier, and which when hit with a rock gave out a clear ring — not high-pitched, but a low note reminiscent of a bell-buoy.

Anyone looking for this stone should have no trouble. It lies in wild surroundings, away from cottages. The narrow, swampy canoe passage comes in off the lake immediately south of Swift (not Little) Current, where the highway and railway make their first short jump from the mainland toward Manitoulin Island. Looking across the last mile-long stretch of water as one heads south, having rounded the final point from the Swift Current bridge, appears what seems to be a flat, low meadow. Closer inspection reveals the old canoe passage — so narrow and shallow in September, 1967, that the canoe had to be dragged through empty. This little ditch is short, and from its southern end the rock, *La Cloche,* lies on an open rocky beach to the east of the reedy swamp, which was once the voyageurs' way through. It lines up at an angle of about 45° to the axis of the "ditch", and at a distance of 100-150 yards from its southern end. Two or three smaller "cloches" are nearby. Seekers should perhaps be warned that the name, "Bell Rock", is locally and mistakenly given to a large block of *granite* half a mile west, which of course does not ring.

The North Channel runs as far as Sault Ste. Marie. In earlier days it was Michilimackinac that was the great depot for reprovisioning and re-routing cargo — south by Lake Michigan, west by Green Bay and the Fox, northwest by Lake Superior. Canoes for the Northwest, leaving Mackinac, doubled back to Sault Ste. Marie by way of *Le Détour,* inside Drummond Island. Once the Athabasca trade became properly established, the Montreal canoes bypassed Michilimackinac and reprovisioned at Sault Ste. Marie. The North West Company built a small canal, just capable of carrying Montreal canoes, past the St. Mary's rapids. One of the first tiny locks is preserved in its original place, today.

Lake Superior

The canoe route around Lake Superior simply keeps within reasonable distance of shore. Lake Superior's north shore, however, in places is cut up by deep bays — one or two of them fifty miles long — and in stretches, particularly around Michipicoten, its rocky shore rises for up to 500 feet straight out of the water. Four hundred miles long, Lake Superior is a sea; and its shoreline in exposed sections, barren of a blade of grass for twenty or thirty feet above the water-line, attests to the power of its stormy waves. Paul Fountain writing in the middle of last century has a pertinent paragraph:

> "These lakes (Huron and Superior) ... are subject to sudden squalls ... Unless therefore you have time to reach the land, an upset is inevitable. Consequently it is necessary to creep around the shore; but when a bay ... is come to, the crew, naturally, to save time and labour like to strike straight across from headland to headland. As some of the traverses are not less than twenty miles broad, it is necessary to study the weather and to be an accurate judge of ... probabilities ... For the waves of Huron and Superior are not inferior in size and power to those of the ocean, if indeed, they are not more to be dreaded."

Lake Superior in a sense has a climate of its own, for this great mass of water (1300 feet deep in spots) never really warms up. In June, the month when the voyageurs were crossing it on their way to Grand Portage, there is frequent heavy fog. In July and August, though there is less fog, large areas of hot, bare rock beside such a great volume of cold water set up violent air currents, so that by early afternoon the lake often becomes unsafe for small craft.

There is an understandable human tendency in the journalist to stress the hazards. However, we know that the fur canoes got through regularly, sometimes recording extraordinary speeds. On the other side of the picture, and less well publicized, are the factors which aided the canoes. The ordinary map, for instance, fails to show the detail of the coast line; actually, for more than 100 of the 450 miles from Sault Ste. Marie to Grand Portage, there is a screen of coastal islands. To beat the wind, the voyageurs could paddle at night; and they certainly would be on the water by three in the morning. The only risk on a night journey would be in an area of shallow, submerged rock, as in the *Pays Plat* west of the mouth of the Nipigon.

As the transport system became better organized, the North West Company put sail vessels on Lake Superior. These took on board about half the cargo of each canoe, allowing it more freeboard and relatively more power. A bark canoe, without a spine, in any case was vulnerable to waves. The waves of Lake Superior, on account of the Lake's great depth, are surprisingly long. We found, ourselves, that an ordinary two-man, 17-foot canoe had tolerances for six- and four-foot waves, respectively from stern and beam. As for the high, rocky portions of shore, none was in fact more than ten miles long — and most, less than five. Between them were many little harbours which, even though they forbade a landing, might let a brigade stand out from shore and ride out a squall. The squalls are of extraordinary violence and suddenness. If the wind happened to be favourable, of course, it gave the emaciated voyageurs, who might be at the end of six weeks' gruelling travel, the supreme joy of a lazy day, for the canoe was equipped with a big square sail which probably could allow it to stay on the water for up to a speed of eight or ten knots. The voyageurs spoke of the wind, so vital to their progress, as *La Vieille*, the Old Lady. In a canoe on a big lake, during one of *La Vieille's* tantrums the party takes to shore. In the fur trade days, the packs would be laid out to dry and air, while the voyageurs relaxed. They termed this being *dégradé*. When *La Vieille* smiled again, the rested crew pressed on to make up the time lost. In the various accounts of travel on Lake Superior, there is a surprising, general agreement on the proportion of time a canoe is pinned down by unmanageable wind during the two best months, July and August: it works out to one day in three. In less favourable months, the proportion might be half and half.

The canoe trip around Lake Superior's north shore is exciting and rewarding, but should not be undertaken by canoeists lacking in experience and weather-sense. Because of the risk of sudden squalls, and the physical dimensions of everything associated with Lake Superior, a party of more than two canoes might have trouble keeping contact. Our rule of thumb was always to be within four miles of shore.

While the total absence of portages on a lake always has the effect of lessening one's sense of contact with historic predecessors, on Lake Superior this is offset to a degree by the procession of geographical names, originating with the voyageurs. On such a journey, with half the day's course in view, a headland or piece of coast sometimes from far off assumes the form of an animal or object. Many voyageur names happily have been officially adopted by the Canadian Permanent Committee on Geographical Names — names like Pancake Bay, Otter Head, Bottle Point, Roche Debout, Les Mammelles, Sleeping Giant, and Pie Island, all easily recognizable features. Approaching it from the north ten miles away, a modern voyageur having a passing acquaintance with Rabelais and a voyageur's low mind would be amused to divine the meaning of Cape Gargantua.

The *canots de maître* ended their journey at Grand Portage or Fort William, which brings this chapter to an end. There remains only to discuss certain routes east of Grand Portage, which were feeders, connections, or experiments.

Lakes Huron & Superior Reading

3	II 104-31, 190-237	**15**	85-92	**27**	43-45
14	107-16	**20**	33-37, 59-67, 198-207, 230-5	**28**	II 145-48

Branch Routes

These lesser routes can be more conveniently grouped by region than by function: (i) connections between James Bay and the St. Lawrence-Great Lakes system; (ii) heavy-freight routes in the Great Lakes area.

Since the James Bay and the St. Lawrence basins represented the respective bailiwicks of two rival fur companies, one would hardly look for trade routes between them. There is Father Albanel's 1672 route via Lake St. John, Lake Mistassini, and the Rupert River — a route still paddled. The Eastmain River also made connections with Lake Mistassini. From 1821-83 the Hudson's Bay Co. post on Lake Timiskaming was

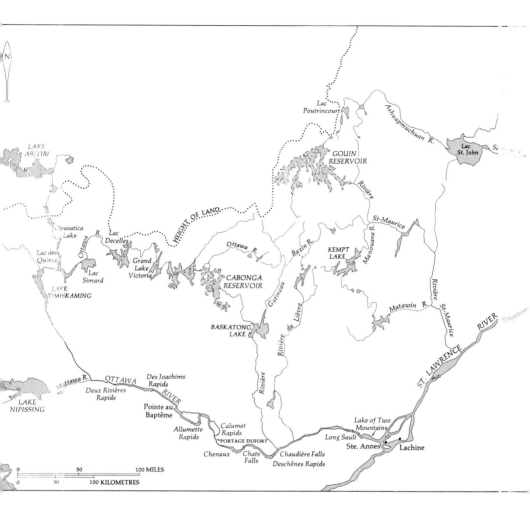

The Ottawa, and its clustered sources

serviced from Moose Factory. The canoe route followed was that used earlier by French military raiding parties, such as De Troyes, passing over the divide by way of Lac des Quinze on the Ottawa and Lake Abitibi. This is the best connecting route between the Ottawa and James Bay suitable for the carriage of freight — a narrow, winding corridor which, on its course over the height of land barely achieves keeping beneath the 1000-foot contour line. The Canadian artist, Arthur Heming, accompanied a local fur brigade between Abitibi Post and Timiskaming over this route in 1901. An abridgement of his story appeared in the Summer 1967 *Beaver*, with his vivid sketches.

In this same general area, but not actually crossing the divide, was a temporary detour caused by the Iroquois domination of the Ottawa Valley in the seventeenth century. Huron and Nipissing middlemen bringing their furs down to Montreal in those perilous years, when they came to the mouth of the Mattawa turned *left* up the Ottawa. The sources of the Ottawa, St. Maurice, and Saguenay lie close enough together to let small canoes get through to Montreal this roundabout way. This route was not important for volume of traffic, nor length of service; it merely filled an emergency for a few years, till the Iroquois menace was over. (See accompanying map)

The second most common link between the St. Lawrence system and James Bay was by way of the Michipicoten and Missinaibi Rivers. It was too rugged for the carriage of heavy freight, and served more as an express route between Lake Superior and Moose Factory. The voyageurs used the Magpie River, and Wawa and Hawk Lakes, to get up the steep gradient at its Lake Superior end. A party which made the trip in 1967 describes the route as having become spoiled in many of its previously attractive sections.

Dr. S. E. Dawson's admirable *"The Saint Lawrence Basin"* covers in detail the twenty-or-so main "gateways" connecting the St. Lawrence with its neighbouring basins. Though these were all once canoe routes, not all were trade routes.

Within the Great Lakes area, bateaux, sail and steam vessels came to supplement canoe freighting, but for a variety of reasons which lie outside the scope of this book, growth of freight traffic on the Great Lakes did not seem appreciably to relieve the very expensive freight-canoe traffic carried on the original Ottawa-Nipissing route down to 1821. An Indian short-cut between Lake Huron and Lake Ontario lay through Lake Simcoe. Among its well-known travellers were Etienne Brûlé, Alexander Henry,

Maj. J. H. Lefroy, and Dr. Bigsby. Such a small proportion of the mileage was by water that its classification as a canoe route is marginal. As a freight route (via Yonge St.) it failed to achieve serious use — except possibly between 1815-21. The Chamber of Commerce at Barrie in the nineteen fifties restored what was still traceable of the Nine Mile Portage which led from Lake Simcoe to Willow Creek, a tributary of the Nottawasaga and the water-link to Georgian Bay.

Chapter VI

Montrealers' Mainline:
2 / Grand Portage to Fort Chipewyan

From Grand Portage and Fort William on Lake Superior, two routes started off westward, meeting at Lac la Croix, respectively after 150 and 230 miles. Both routes crossed the height of land separating St. Lawrence and Hudson Bay waters, at about 900 feet above the level of Lake Superior. This was the steepest part of the voyageurs' whole route to Fort Chipewyan. Impressively, from Grand Portage the Hudson Bay height of land lies only fifty miles back from Lake Superior, involving (for freight canoes) a very high average rate of climb, eighteen feet to the mile. On the Fort William route the average gradient, still steep, is half this. From Lake Superior westward, North Canoes were used.

The Border Lakes Route to Lac la Croix

The Border Lakes route follows the international boundary to Lake of the Woods. More accurately, the international boundary follows the lakes, for the Commissioners in settling on a line through this complicated country adopted as their working principle the "customary waterway" of the early voyageurs. The Pigeon River basically is the boundary east of the divide. However, the Pigeon because of a series of cascades and canyons, is totally unnavigable near its mouth. From a shallow bay ten miles south of its mouth, a nine-mile Indian trail, the Grand Portage, cuts across to a point on the Pigeon above the unnavigable section, where Fort Charlotte once lay. Alexander Mackenzie perhaps unconsciously reveals the eighteenth-century entrepreneur in describing the voyageurs' role on this portage:

> "When they arrived at the Grand Portage . . . each of them has to carry eight packages of such goods and provisions as are necessary for the interior country. This is a labour which cattle cannot conveniently perform in summer, as both horses and oxen were tried by the Company without success."

A photograph taken from inside the palisade at reconstructed Grand Portage fort, looking out past one of the bastions toward Lake Superior. On the right is Grand Portage Island, on the left, Hat Point, around which the voyageurs at the end of their journey from Montreal would stop to wash, shave, and don a clean shirt and *ceinture fléchée*.

Grand Portage Post, reconstructed with stockade and bastions, is worth visiting. The nine-mile trail, resembling just a lumber road, is in good condition. By treaty all these portages along the border lakes route were internationalized, so that today, without hindrance or formality, anyone can camp or lunch in either country as he paddles the international boundary. Identification with the past is particularly vivid when, along the portages, it is realized that a towering pine beside the path could have been there as a seedling when Mackenzie or Thompson passed. Thanks to the Superior National Forest Area on one side of the line, and Ontario's Quetico Park on the other, there is some assurance that these lakes and portages will be preserved for posterity in their original state, looking as La Vérendrye saw them.

Well, almost. Or for as far west as Crooked Lake. At Curtain Falls below Crooked Lake, American authorities have not succeeded in ejecting an unsightly fishing resort. Denied of service by plane, owing to an air-space regulation, this resort supplies itself by surface transportation, and appears to have taken matters into its own hands. In the process, the two historic portages, Curtain Falls and Bottle, leading to Lac la Croix have been bulldozed beyond hope of return to their natural state. The late-lamented Bottle Portage lies in Ontario's Quetico Park, a commentary on international amity in this region.

The Grand Portage route immediately west of Lake Superior was dramatic, different, and exciting. The early diarists made it rich in record: especially Mackenzie, both Henry's, Harmon and Bigsby. The twenty-nine portages, totalling fifteen miles, along the 150 miles to Lac la Croix are easy to locate — except for a couple on the Pigeon River.

Attention is directed to only one particular stretch approaching the height of land, where the drainage is more intricate, and the original route, changed. The continental divide lies between South and North Lakes, on an easy, 700-yard portage. South Lake does not flow directly into the Pigeon River, as is often assumed, but into a tributary, the Arrow River. The highest Pigeon River lake which the route crosses actually is Mountain Lake, which is 125 feet *higher* than Rose Lake to its west. The original Indian route passed from Mountain to Rose Lake by way of Arrow Lake. The Indian portage to Arrow Lake must have been rough and long, for this was one of the very few places where the fur traders left the Indian route and made portages of their own. By the Petit Neuf (now Watape) Portage they cut directly from Mountain Lake over to Watape Lake, which is in

Arrow Lake drainage. At the Petit Détroit (blocked by beaver in 1954) through which Watape Lake flows into Rove Lake, the unwitting voyageur, not aware of having switched to a new river, may be bewildered by seeing the current now flowing with him and be worried that he is heading back for Lake Superior. From Rove Lake, Grand Neuf (now Long) Portage led to Rose Lake. The portage is rough, and at its western end leads into an abandoned rail-bed once used for logging. Mackenzie sums up this new route:

> *"In the centre of [Mountain] Lake, and to the [north] is i'.e Old Road, by which I never passed, but an adequate notion may be formed of it from the road I am going to describe, and which is universally preferred. This is first the Small New Portage . . .; the whole is then embarked on a narrow line of water [Watape and Rove Lakes] that meanders South West about two miles and a half. It is necessary to unload [at the Petit Détroit], for the length of the canoe, and then proceed West half a mile, to the New Grand Portage, which is three thousand one hundred paces in length, and over very rough ground, which requires the utmost exertions of the men and frequently lames them. From hence they approach Rose Lake, the portage of that name being opposite to the junction of the road from Mountain Lake."*

In this context of the Quetico and Minnesota area, Canadians should salute certain American authorities, voluntary associations, and dedicated individuals for their generous and imaginative contribution to preservation. The story of the Canadian fur trade owes a great debt too, for research and general popularization, to the Minnesota Historical Society.

Now having proceeded along the international border to Lac la Croix, we must go back to follow the alternative route (from Fort William), in operation after 1803. Lac la Croix probably got its name from being a crossing of routes; for there was a further split below this lake.

Border Lakes Route Reading

1	272-9	14	125-28	20	239-41	27	46-65
3	239-312	15	92-104	22	81-84	29	3-9
9	8-26	18	14-20	24	II 110-13, 129-31	56	1-16

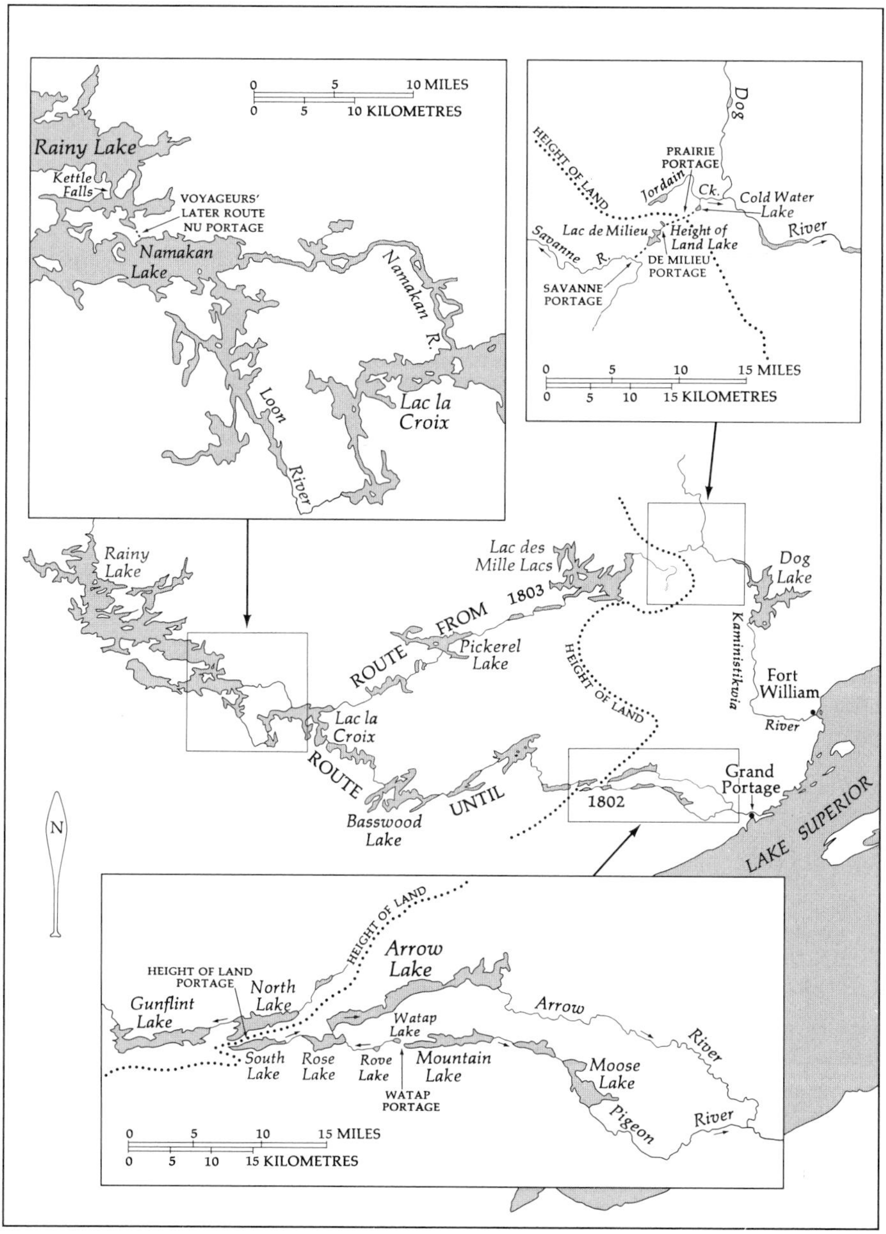
0 5 10 MILES
0 5 10 KILOMETRES
Rainy Lake
Kettle Falls
VOYAGEURS' LATER ROUTE NU PORTAGE
Namakan Lake
Namakan R.
Loon River
Lac la Croix
HEIGHT OF LAND
PRAIRIE PORTAGE
Dog
Jordain Ck.
Cold Water Lake
River
Lac de Milieu
Height of Land Lake
Savanne R.
DE MILIEU PORTAGE
SAVANNE PORTAGE
0 5 10 15 MILES
0 5 10 15 KILOMETRES
Rainy Lake
Lac des Mille Lacs
ROUTE FROM 1803
Pickerel Lake
Dog Lake
Kaministikwia
HEIGHT OF LAND
Fort William
River
Lac la Croix
ROUTE UNTIL 1802
Basswood Lake
Grand Portage
LAKE SUPERIOR
N
HEIGHT OF LAND
Arrow Lake
HEIGHT OF LAND PORTAGE
North Lake
Gunflint Lake
South Lake
Rose Lake
Watap Lake
Rove Lake
Mountain Lake
WATAP PORTAGE
Arrow River
Moose Lake
Pigeon River
0 5 10 15 MILES
0 5 10 15 KILOMETRES

Lake Superior to Rainy Lake

The Kam — Dog — Maligne Route

The first route travelled by a trader from Lake Superior to Rainy Lake was the one starting up the Kaministikwia; de Noyon who in 1688 had preceded La Vérendrye in reaching Rainy Lake went this way. La Vérendrye "discovered" the Indians' Grand Portage route just described, and since it was easier, it came into use; the Kaministikwia route was abandoned and forgotten. It was not till several years later, when it became apparent that the Americans proposed to levy duties at Grand Portage, that the original route was re-explored by Roderic Mackenzie. By 1803, the shift from Grand Portage Post to Fort William was completed, and the Kam-Dog Route came back into use. It was employed by the North West Company for heavy freight for only eighteen years. Express canoes continued to use this route till the railway came.

The actual North West Company establishment at Fort William stood on what is now the C.P.R. railway yard, on the north bank of the Kaministikwia, opposite where the middle channel broke off. From here the canoes headed up the river. For thirty miles up the Kaministikwia there are shallow riffles and fast water, and one *décharge, des Paresseux*. The first portage is long and bad, called Mountain Portage, passing 120 ft. Kakabeka Falls. Curiously, only one account makes specifically clear on which side of the river the old portage lay. Prof. William Keating, narrator to Major Long's 1823 American surveying expedition, states:

> "About a quarter of a mile below the fall there is a sort of cove in the right [i.e. west] bank; it is about 300 yds. in diameter, and bounded by steep bluffs on all sides, except in front, where the river passes. The Portage road, which is about three quarters of a mile long terminates at this place; a descent to the level of the water having been made by the Northwest Company."

This path, when I last saw it, was much overgrown with scrub, and easier to locate while the trees were leafless. There is only one spot along the encircling bluff where a man can get up or down, and this may be the only definitely relocatable portion of the portage; it does, however, give the general line.

The Northwest Company, to save time and toil, conveyed most of the packs by road from Fort William to a depot above Kakabeka Falls. From here the ascending voyageurs were faced by a mean river which, in the first 23 miles above the Mountain Portage, drops at nearly ten feet to the mile, including seven portages and two *décharges*. The descriptions of this section by many travellers going up or downstream, and at varying water levels are confusing, even in nomenclature. I favour for these reasons accepting as most authoritative the naming and data of H. Y. Hind who officially surveyed the route nearest the close of its period of regular use. In Appendix I, page 399, of his *Narrative of the Canadian Red River Exploring Expedition of 1857,* Hind gives the length of each portage, its distance from Lake Superior, and its elevation. Most of these rapids have been blasted out for a hydro development, leaving little sense of the past on this section of the river. The portages left are reasonably easy of approach by highway for as far as Little Dog Lake and the start of the Great Dog Portage —required to get past a series of bad rapids below Dog Lake.

This curious name, dominating the area, arose from an Indian effigy of an out-size dog lying at the site of a remarkable view from a 400-foot hill, looking down the Kaministikwia valley. Professor Kenneth Dawson has located the dog effigy and excavated it, so that its exact site is known. It is on the crest of the hill, above the hydro standpipe. The voyageurs portaged by way of a long and obvious draw, a little to the east. The draw afforded them a much easier gradient to attack a 500-foot hill. The portage trail is marked on the first edition (1939) of map 52 $\frac{A}{N.W.}$; scale: 1 inch = 2 miles. The map shows the portage as 1⅔ miles, matching Hind's survey.

Because of beaver flooding at the south end, the portage can be picked up better from Great Dog Lake, where it started from a small beach, in fact the only beach along that part of the lake. Palliser identifies this as lying three and a half miles east of the lake's outlet. It is due north of the east shore of Little Dog Lake. From the beach, the portage trail leads up a steep, narrow gully to a cleared area, then heads due south, soon feeding into an old dirt road which from its direction almost certainly lies on the original portage path. After crossing the main road, the trail becomes a car-width forest road continuing south down the long draw described above. Beaver by 1975 had made it impossible to follow the trail beyond the foot of the big hill. However, anyone wishing to retrace the walkable upper two-thirds of the portage can locate it with the help of Air Photo A22904-38. The best approach is to park a car where the portage trail crosses the main road and find one's way on foot down to the beach on Great Dog Lake.

From a coloured print of the lower landing of the portage on Little Dog Lake in Hind's *Narrative,* the landing was fairly close to the foot of the rapids — in fact, probably right beside the foot of the present hydro tailrace.

Over the Dog — Savanne Divide

Launched on Dog Lake, the voyageurs had fifty miles with negligible portaging as they crossed Dog Lake and paddled up marshy, winding Dog River, Jordain Creek, and Cold Water Creek, to Cold Water Lake. Cold Water Lake, spring-fed by its clustering hills, deserves its name — even in the height of summer. From the lake's westward extremity began the series of three long, hilly, boggy portages which necessitated the North West Company paying voyageurs a bonus for coming this way. In succesion were the three-mile *Prairie* Portage, the half-mile *de Milieu* Portage, and the mile-and-a-half *Savanne* Portage, separated respectively by Height of Land Lake and Lac de Milieu.

Since, on local inquiry in 1967, I discovered that the last two parties to attempt to cross the divide had been turned back, unable to find the trail, I approached this six-mile section from opposite ends, without crossing it. I eventually succeeded in tracking down a senior executive of the paper company holding these limits, who kept a canoe on Lac de Milieu (now called Savanne L.). This informed and enthusiastic gentleman* wrote the following up-to-date and first-hand information, which, in the decade of the apparent closing of this historic divide to canoe travel, is appropriate to record here. He said in a letter dated September, 1967:

> "I have attempted to follow the Prairie Portage eastward (to Cold
> Water Lake), but I ran into difficulty. Several trails had been used
> at different times to avoid the frequent blow-downs in the over-
> mature spruce swamp . . . I followed one out of the swamp and
> into the jack pine upland, but lost it there and had to turn back.
> I have used the Portage de Milieu as recently as 1965 and found
> it very well marked. The trail is worn three to six inches deep in
> the softer soils and is only overgrown by shrubs in two places.
> Old blazes are still frequent . . . At Lac de Milieu the portage starts
> on a point north of the bay that would be the natural starting
> place . . . From here it crosses a flat then climbs over a jack pine
> ridge. Beyond this . . . [overgrown, swampy] it is necessary to
> follow it carefully. Then the trail descends gradually through a
> poplar stand to the bog surrounding the unnamed (Height of Land)
> lake at its end. At the west side of Lac de Milieu (Savanne L.) there
> is a broad bog stretching to the highway through which fire has
> run at least once, apparently destroying all signs of the Savanne

---

* Mr. Stanley Losee of the Abitibi Pulp and Paper Co.

Portage. Even aerial photos fail to yield any clues . . . When I saw it last in 1964, the creek had a beaver dam; . . . and between low water, driftwood, and windfalls, was quite impassable."

The route from the Savanne River to Lac la Croix is classifiable with the Border Lakes section, that is, still unchanged and in use. The canoes passed from the slow, winding Savanne River into beautiful, island-studded Lac des Mille Lacs. It should be stressed that from Lac des Mille Lacs the canoe route did not follow the Seine, but passed over a divide at quarter-mile Baril Portage directly into another river, the Pickerel. From the west end of Pickerel Lake the voyageurs crossed by the Pickerel and Deux Rivières portages into Sturgeon Lake, thence down the Maligne River to Lac la Croix. The Seine, though more direct, was too rough for canoe-freighting. The whole route from Lac des Mille Lacs to la Croix is attractive, unspoiled, and easily accessible.

Fort William — Dog R. — Lac la Croix Reading

1	280-90	15	195-99	23	35-38
14	118-24, 183	18	113-14	24	II 136-47, 174

Lac la Croix to Rainy Lake

Lac la Croix flows out in two streams of about the same size, by way of Loon River and Namakan River. The former was used as the canoe route during the whole period of heavy freight traffic under the North West Company — as is evidenced by the international boundary. Loon River passes out of Lac la Croix's south-west tip and goes through Vermilion and Sand Point Lakes, before flowing into Lake Namakan. Its three old portages have now become marine railways, constructed to assist the heavy motor-boat traffic of this river.

The Namakan River leaves Lac la Croix from a bay in its north shore, and, striking much more directly for Namakan Lake, takes the drop in fewer miles. Frances Simpson, the Governor's eighteen-year-old bride, in the 1830 Diary of her trip west explains how it was that the Namakan River became the preferred route after this date. Her entry for May 31st reveals a good grasp of geography and logistics:

"Descended the river Michan [an alternative spelling], which is not the usual route, but much shorter; yet seldom or never passed by Whites, being considered dangerous . . . Mr. Simpson therefore to

ascertain its state, mounted its current last Fall, in low water, and considers it not only a practicable, but a safe route in any state, either for Boats or Canoes."

The Namakan today is also wilder and better canoeing.

Between Namakan and Rainy Lakes, the complicated water pattern of this country offers still another option, for Namakan Lake, again, has two outlets. Originally the canoe route followed Lake Namakan to its western end, portaging around Kettle Falls. However, this meant doubling back eastward again and was riskier, because of volume of water and the falls. A new route was developed — it would seem soon after the North West Company was organized — following a smaller water course which flows out from a more easterly point on Lake Namakan and heads directly for Rainy Lake, passing just below Kettle Falls. Two small portages past rocky corners were involved, known as *Nu* (also *Neuf, Nouvelle*).

Rainy Lake is still attractive, though no longer wild. At its lower end are the twin paper-mill cities of Fort Frances, Ontario, and International Falls, Minnesota. The Chaudière Falls, a short distance down Rainy River from the outlet of Rainy Lake, are now dammed, above which the river is blocked by log booms. The easiest way through the urban area is to paddle down the river on the Ontario side a couple of miles and take out at some obvious point, hiring a truck to get below the dam. The site of Fort La Pluie lies a little below the dam on the north shore; it has been reconstructed and is worth visiting. Here the special brigade of "pork eaters" came out from Grand Portage to meet and exchange packs with the Athabasca Brigade, allowing the latter a head start to beat the frost on their way home.

Rainy River and Lake of the Woods

Rainy River, about 80 miles in length, flows from Rainy Lake to the south-east corner of Lake of the Woods, at which point the international boundary ceases to follow the old canoe route. Rainy River valley was once a part of geological Lake Agassiz, and the river's shores for the most part consist of good alluvial soil. On the canoe route westward, through the rock country of the Shield, the early travellers (nearly all from Britain or France) had seen nothing like this since leaving the lower Ottawa. M'Gillivray said it was "reckoned the most beautifull River in the North, a preference which it richly deserves". Other diarists went into similar nostalgic raptures. Actually the Rainy River is monotonous.

Lake of the Woods, with its thousands of islands and the presence once more of clean granite, is attractive — particularly in its wilder, middle section. A look at a map of the lake shows the need for most diligent navigating; in fact, the voyageurs seem to have got lost more often in Lake of the Woods than in all the other miles of their long voyage put together. Aulneau was a peninsula by a thread (now a canal cuts its neck); and it would seem that this more protected eastern route — even including portaging — would be better. However, the voyageurs, from the mouth of the Rainy River, made the *Grande Traverse* over an open horizon to Bigsby Island, passing east or west of this, as *La Vieille* dictated. They were heading to graze the western tip of Aulneau Peninsula, a more direct, nearly north-south route, only seventy-five miles long. At the tip of the Peninsula they met an oddity: a portage in the middle of the lake, as it were — at least at low water. Today the dam at the foot of the lake ensures sufficient draft here for small craft at all seasons.

The Voyageurs' Rat Portage

Lake of the Woods empties into the Winnipeg River by three channels, the Eastern and Western Outlets (now in or bordering on Kenora) and a small outlet from Portage Bay in the town of Keewatin, three miles west. Each channel had its own portage — which did the voyageurs use? According to the journals of Alexander Henry, Alexander Mackenzie and Dr. J. J. Bigsby, all of whom were describing what they saw between 1763 and 1822, the voyageurs here carried for not more than fifty paces. This pinpoints with little doubt where the voyageurs portaged — though the descriptions of later travellers indicate that it was not used after 1821.

Dr. Bigsby, Secretary of the Boundary Commission, was at Rat Portage in 1822, the year after the voyageurs stopped coming through with heavy freight. In his *The Shoe and Canoe* is a map by David Thompson, at that time "Astronomer" to the Commission; this map confirms that the voyageurs used the Keewatin Channel, portaging well to the west of Portage Bay (see map p. 63 of this text). On page 303, Vol. II of his book, Dr. Bigsby elaborates with some precision:

> "The Rat Portage, we reach by a narrow cul-de-sac, 600 yards long, ending in a grassy swamp, the portage lying between two eminences. . . . This cul-de-sac is 120 yards broad at the portage, and is made offensive and foul by dead insects . . . and the plague of mosquitoes.

An early sketch on Great Dog Portage, showing a North Canoe being portaged in
its upright position with the keel on the shoulder. Normally two men portaged a
North Canoe; the third man shown here may have been added because the Great
Dog Portage is steep and long.

The portage is a neck of land fifty paces across, between the dirty cove in the Lake and a magnificent sheet of water". (Darlington Bay). His description and other features he mentions would point to the cul-de-sac as being the narrow, western arm of Portage Bay, which broadened into Mink Bay. This dead-end bay spilled only at high water, accounting for its stagnancy. The voyageurs' portage would thus appear now to lie buried under the C.P.R. tracks; and the "neck of land 50 paces across" to have become the bedrock foundation for the railway embankment crossing this gap.

The obvious advantage which the Keewatin site gave the heavily laden voyageurs was that it was so short. After 1821 when (with the big fur company merger) the main freight route was shifted to York Factory and travellers passing Rat Portage were no longer each carrying 180 pounds, a more direct route but involving a longer carry of 150 paces seems to have been in use. It, too, led from Portage Bay, but instead of swinging westward via Bigsby's cul-de-sac it kept straight north. The portage trail appears to be still traceable in a wagon track leading down an easy gradient just west of a small water course. A burned-down flour mill close to the C.P.R. tracks marked its upper end in 1975. This was probably the portage used by Keating and Alexander Henry the Younger. From the mid-1830's, with the establishment of the trading post and nearby village of Rat Portage on the Eastern Outlet, all travellers — including Lefroy, Kane, Hind, Palliser and Dawson — used the portage along the shore of the eastern mainland.

Dr. A. P. Coleman in the Fifth Report of the Ontario Bureau of Mines, 1895, and also the International Joint Commission's Final Report on the Lake of the Woods Reference, 1917, describe the industrial development which complicated the natural drainage pattern in the area of Keewatin — including blasting through the rock in the 1880's a whole new channel as a flume to a flour mill which stood near the present boat lift. Dr. Coleman on pp. 168–71 of his above report makes an important historical contribution, in that he was writing within seventy-five years of the abandonment by the voyageurs of the Keewatin site, and so was able to talk with local old-timers who recalled hearing significant details.

The Winnipeg River

The Winnipeg River was unquestionably the grandest and most beautiful river the Montreal Northmen saw on their whole journey from Lake Superior to Lake Athabasca. Running often through tortured rock and dropping fast, it was a river of spectacular rapids and falls. Twenty-six historic carrying places marked its course. A metropolis in flat country, and less than 100 miles away, has now reached out and harnessed the Winnipeg's wild grandeur: of its 272 feet of drop within Manitoba, six power dams take up all but five feet. Nevertheless, except at the power sites themselves, the Winnipeg is still attractive and moderately wild; the water, considering the statistics, is still surprisingly alive; this is especially so at high water when dams are spilling. Eight (a third) of its historic portages are still traceable. For thirty-three miles below Kenora, to about where the C.N.R. crosses at Minaki we have a fresh, free-flowing, often dancing river, still unspoiled. Its uniform and modest drop of only ten feet has spared this section. Eight miles down, the Dalles (or "eave trough", a name encountered earlier) has been slightly modified by having a side channel blasted out to reduce turbulence, so is no longer quite what the up-paddling voyageurs had to contend with in an earlier day.

About ten miles below the Minaki railway bridge, the Winnipeg once followed two separate channels around a big island. Now the Whitedog Dam floods the left channel, while the right channel has been blocked off by rockfill. This north channel is a small "museum" of fur-trade travel, for by luck it was this channel which the voyageurs took. While the south channel lies inundated, the north one preserves in its scant five miles no less than five of the Winnipeg's twenty-six carrying places. The rockfill lets enough water seep through for marginal canoe passage, the flow having been reduced to perhaps a twentieth of its original volume. The portages in 1963, only three years after the building of the dam, were still easily traceable; but with traffic now having forsaken them, the wilderness may already have reclaimed some of the old pathways. The Grand Décharge portage path is now partly buried under the rockfill, but in 1963 could still be located, on the left. With this as mileage zero, the following table summarizes these portages:

Mileage	Name	Side	Paces
0	Grand Décharge	L.	300
2	Terre Jaune	R.	200
2.5	Charette (Petit Rocher)	L.	60
4	Terre Blanche	R.	200
4.5	La Cave	R.	50

Even though the old water-course looks forlorn in its rocky emptiness, a strong sense of identification is possible — through such features as the yellow rock at Terre Jaune, the white clay at Terre Blanche, the dark hollow in the rock beside La Cave rapid, and by the portage paths themselves. These few surviving portages seem to gain significance from the Winnipeg's many celebrated ones which have gone. Anyone passing through today, if too heavily loaded to relish portaging, could visit the Grand Décharge merely from above it, then paddle down the left channel, portage around the right wing of the dam and work up, unloaded, to the other portages from below.

Nineteen more miles of unspoiled river now follow, to Portage de l'Isle (Boundary Falls). The voyageurs' route continued straight ahead here and the portage would appear to have gone, flat and straight, across the low north end of the larger, well-wooded island to its lower landing at a little beach. Alexander Henry the Younger relates a fatal accident which befell one of his crew here in 1800.

> "One of my canoes, to avoid the trouble of making this portage, had not gone many yards when, by some mismanagement of the foreman, the current bore down her bow full upon the shore, against a rock; . . . the canoe was instantly carried out and lost to view amongst the high waves. At length she appeared and stood perpendicular for a moment, when she sank down again, and I then perceived [a] man riding upon a bale of dry goods in the midst of the waves . . . But alas! he sank under a heavy swell, and when the bale arose the man appeared no more . . . The canoe we found flat upon the water, broken in many places."

Henry catalogued his losses: merchandise, tobacco, kettles, shot, guns. In an underwater search program to comb all west-flowing fur-trade rivers, the Minnesota Historical Society, in cooperation with the Royal Ontario Museum, in 1966 investigated the scene of Henry's accident and brought up the biggest haul of submerged fur-trade hardware yet discovered, including much of what Henry had enumerated.

Now the Winnipeg begins to drop faster and, in the sixty miles to the level of Lac du Bonnet, loses 160 feet in altitude. In this section were fourteen of the twenty-six portages — of which only one, at so-called "Sturgeon Falls", is left. The obstructions and portages between Boundary Falls and Lac du Bonnet are here tabulated:

Mileage	Present Obstruction	Historic Name	Historic Portage	Modern Portage
0	Boundary Falls	Portage de l'Isle	(island)	(same)
23	Lamprey Falls	Chute à Jacques	L.	L. or R.
33	Pointe du Bois Dam	Pointe des Bois and Pointe aux Chênes	R.	L.
34	Eight Foot Falls	Rocher Brûlé	R.	(run)
39	Slave Falls Dam	Slave Falls	R.	L.
45	Sturgeon Falls	Barrier Falls	L.	L.
50	Narrows	Grand Rapid (Otter F.)	(ran)	(run)
58	Seven Sisters Dam	Blanche River (7 port.)	various	R.

For the benefit of anyone planning to paddle down the Winnipeg revisiting the old route, a few brief notes are given, in the interest of time, safety, and history. At Boundary Falls the Henry tragedy occurred in a narrow, crooked, and rocky channel lying against the north mainland. A new channel against the south shore has now been blasted out, reducing turbulence and lowering the gradient in both main channels, making them much safer to run. The portage around the Pointe du Bois dam on the left is a bit rough and overgrown, but it saves portaging farther on the other side and through the village main street. "Eight Foot Falls" may sound terrifying to run, but blasting has flattened out the drop, since the days when Rocher Brûlé portage was required. It should be run close to the left bank. The approach to Slave Falls portage is confusing, and also dangerous, since the portage starts close to the sluice-way. Follow the left channel and keep toward the left shore, taking a swing to make a landing in a shallow bay of the island ahead, between the two dams. The right-hand dam marks the sluice-way. From the bay indicated, paths go right and left; the better way is to the left, down to the discharge channel below the east dam.

The local tourist resorts near Sturgeon Falls seem to have succeeded in shifting Barrier Falls two miles downstream, applying this historic name now to a non-falls where a narrow peninsula looking like a barrier juts across the river. The historic Barrier Falls was actually what is now known as Sturgeon Falls. It is to be hoped that this mistake in calling a fast bend in the river "Barrier Falls" will be rectified — especially since Barrier Falls is the last surviving rapid on the Winnipeg. The mileages given in the journals, the drop, and the description of a rapid with a short portage either on the left or else over a big rock in the channel put La Barrière's location beyond question. Five miles down fom La Barrière was Otter Falls, often called Grand Rapid — though what was "grand" about this three-foot drop is mystifying. The Seven Sisters Dam has reduced it to a fast-flowing narrows, locatable also as being a mile above present Pinawa. Going downstream it was usually run; the portage, going upstream, was only a few paces.

Pinawa River versus Blanche River

The Winnipeg River now splits again, making a big island in this section, all the way down to Lac du Bonnet. Branching off to the north is the Pinawa "River", while ahead to the west lay the voyageurs' "Blanche River" — both being simply alternative channels of the Winnipeg. The question arises, why did parties prefer one over the other? The reason is apparent in Mackenzie's comment, "a safe road when the waters are high". All travellers who specifically mentioned taking the turbulent alternative, Blanche River, passed in August or September, when the water was low. Henry Lefroy, heading down the Winnipeg in *June* of 1843, said of the Winnipeg River:

> "One portion of it [Blanche River] is now dangerous in consequence of the water being so high. We left it therefore for a small stream, Pinawa, which there becomes navigable."

This makes quite clear why and when the Pinawa River was used. Alexander Henry, in one of the earliest accounts we have of it, stated:

> "The stream of the Pinawa is shallow, and its bed rocky and broken. The carrying places are eight in number. The mosquitoes were here in such clouds as to prevent us from taking aim at the ducks, of which we might else have shot many."

For modern canoe travel and especially upstream, the Pinawa probably is the better channel; it offers also more in the way of bird life and scenery. There now are two dams, three portages. Only a limited flow of water today goes down the Pinawa Channel.

Returning to the main river at the fork, Seven Sisters dam, a little to the west, has flooded out "Rivière Blanche", so-called says Mackenzie, "from its being, for a considerable length, a succession of falls and cataracts . . . in so short a space that the whole of them are discernible at the same moment". There were seven portages. The river approaching the dam is now diked, while below the dam the diversion lays bare the rocky bed of the two lowest rapids. To avoid a much longer portage through the village, we portaged around the right of the dam, putting into the blasted-out spillway and hoping there would be no spillage till we were a hundred yards or so down stream. The river below this is dull. Near the foot of Lac du Bonnet, the alternative course via Pinawa Channel comes in.

The least said about the section of the Winnipeg below Lac du Bonnet the better. Three dams, McArthur Dam (24 ft.), Great Falls Dam (60 ft.), and Pine Falls Dam (39 ft.) have between them taken up all the drop, and flooded out the old portages: Bonnet, Cap de Bonnet, Big Bonnet, Petit Rocher, Terre Blanche, Eaux Qui Remuent (Silver Falls), and Pine. Between the dams are dikes and desolation. Below Pine Falls Dam, the river comes to life again for a couple of miles, even showing the odd splash of free, white water; but log booms and the town of Pine Falls soon dissipate any historical illusions. Fort Alexander at Bas de la Rivière, the Northwesters' easternmost pemmican post, probably today looks very little different from what it did — except that there are old cars and stoves now in the garbage along the river banks. Beyond lies Lake Winnipeg; and the Winnipeg River broadens out, with the low, misty line of Elk Island and the long southern (Victoria Beach) peninsula guarding its mouth.

Rat Portage and Winnipeg R. Reading

9	I 26-36	20	244-45	27	65-67
14	128-33	23	43-45	29	8-10
15	104-08	24	II 99-108, 144-47	54	3rd instalm't

Lake Winnipeg

To get the feel of this body of water let us ride with Alexander Henry's (the younger) Saskatchewan Brigade in 1808 and sample a few of his summer-time adventures. He was a Northwester but starting from Red River, the Hudson's Bay Company's normal southern approach, instead of from the mouth of the Winnipeg.

> *Aug. 12th.* We hoisted sail and kept on till two o'clock, when there was every appearance of a squall from the S.W. We had some difficulty in landing, as the rain fell in torrents and the wind blew a gale.
>
> *Aug. 13th.* Long before day we were on the water ... The wind blew a gale; however, we coasted along in the reeds and rushes. We made a long traverse to the mainland, where the shore was so steep and rocky that we could find no place to put ashore, and were obliged to push on in the dark. In a short time the wind rose dead

ahead from the W., and the swell increased. Our position was decidedly unpleasant; the sea dashed with great violence against the rocks, the night was extremely dark, and the wind seemed to be increasing. Anxious to find any place to land we crept on as near the shore as the surge would permit till, having shipped a great quantity of water, we discovered a small cove . . .

Aug. 15th. At eleven o'clock [a.m.] everything was dry, and the wind had abated: we hoisted sail . . . a sudden squall from the N.W. obliged us to put shore at l'Isle d'Encampement where we were detained until three o'clock . . . We kept on with double-reefed sail until nine o'clock, when we camped on a fine sandy beach. We soon had a terrible squall . . . My tent was blown down and we passed a wretched night, wet to the skin.

Aug. 16th. At four o'clock we loaded, and with great difficulty we got around the reef. As the wind continued to blow hard, we shipped much water . . . The swell was so high that, in rounding a point, we nearly filled several times . . . We all got wet to the skin, and our baggage was completely soaked.

Aug. 17th. The swell occasioned by the late gale still ran high. After much trouble in loading, we embarked and stood out on the traverse for the Tête aux Pichaux . . . The wind increased to such a degree in rounding this point, and the sea ran so high while we were under sail, that at intervals we lost sight of the masts of the canoes not more than 30 yards distant; . . . We soon found that our canoe could not stand it much longer, as we shipped great quantities of water; and night coming on, we determined to run ashore, at the risk of breaking our canoe . . . We ran in close to shore and . . . put her about, and kept her stern foremost. Almost every swell washed over her, and as soon as we could find bottom all hands jumped overboard, each taking a load ashore . . . We hauled her up with some difficulty, and camped for the night, during which the wind continued to blow with great violence."

The dates August 18 and 19 in the diary continue to be sprinkled with such phrases as: "increased to such a gale that we could scarcely carry two feet of sail"; "the wind was too strong for us to attempt to round the point". They passed "two kegs of high wine" and "two newly broken paddles" which he suspected to be "part of a cargo that had been wrecked in the late gales". Finally on the 20th, a week out of Red River, Henry entered the mouth of the Saskatchewan River.

No canoe course lay *across* Lake Winnipeg — in any direction. This is probably the worst lake in Canada for small craft. Apart from its dimensions (280 miles long, with a maximum width of 65 miles), it is in the belt of strong prairie winds, and is incredibly shallow — only ten or twelve feet for much of its surface, with an average depth of perhaps no more than thirty or forty feet. The shallowness makes for steep-fronted, choppy waves, not the long swells of Lake Superior. However, the lake is wasp-waisted, with a narrows only two miles across, a third of the way up. There were four canoe routes around its shores: two routes the length of the lake, and one across each end. The Hudson's Bay Co. route around Lake Winnipeg's north end was described in the last chapter. The Hudson's Bay Company and North West Company routes down the lake crossed in the middle at the Narrows. The North West Company brigades leaving Bas de la Rivière turned north up the east coast of the lake; they made two short traverses, to and off Black Island, to avoid detouring around the deep bay to its east; they went through Loon Narrows, separating an island from a prominent point 20 miles south of the Narrows. At the Narrows (Dog's Head) they crossed to the west side, Henry's route. From here the Nor'westers could not always coast safely close to shore, but had to make a total of ten traverses, across mouths of deep bays or between islands — as comes out clearly in Henry's journal. The minimum distance is five miles (an hour's paddling in bad weather), though to save miles the voyageurs often preferred to make longer, riskier traverses. Two exposed points on the way were significantly named "*Maligne*". Just as they neared safety, a fifty-mile detour was required, where Long Point jutted 25 miles into the lake. This, the voyageurs knew as *Le Détour*, and rounding it was often dangerous. If caught in a squall coming up the Lake, the brigade could land, but the shoreline on the west side was largely inhospitable.

Even though the Hudson's Bay Company York Boats were more seaworthy, could carry more sail, and chance a longer traverse, they too kept to the coast. From Fort Garry past the mouth of the Red River, their course lay up the west side of the Lake. Normally they passed through Grassy Narrows, behind Hecla Island. From Grindstone Point they made the traverse to the east shore, taking advantage of a course which offered two or three little islands in mid-traverse. They then held to the east shore of the Lake, which necessitated no risky traverses such as on the west shore.

The short southern route from the mouth of the Red River across to the mouth of the Winnipeg River took advantage of the low spot in the narrow neck just south of Victoria Beach; at low water they dragged their boats over, or portaged canoes; at high water they floated over the neck, from which the course lay straight across Traverse Bay to the mouth of the Winnipeg River.

Lake Winnipeg is neither canoe country nor vacation land. Apart from the hazards from its wind and waves, the algal content in already murky waters makes the water filthy in some of its bays. If, despite this, anyone is determined to paddle Lake Winnipeg, he is advised to start at Norway House (approachable by air) and follow the east coast to Pine Falls. The east coast of the Lake marks the edge of the Shield; and for half the distance there are attractive stretches, with granite points and islands, interspersed with sandy bays. On the other hand, for miles the storm-driven surf has undercut the unstable alluvial shoreline, causing a desolate tangle of fallen trees. There are also miles of sand beaches — whose only redeeming feature is the opportunity they offer for lining a canoe in a contrary wind. Lake Winnipeg makes for a *restful* canoe trip, since canoes are driven off the water most afternoons by two. With an early camp, it is easier to be ready to rise next morning before dawn, to catch the one calm period of the day.

Lake Winnipeg Reading

1	251-57	14	134, 143-46	27	70-72
9	I 36-39, II 453-60	20	250-57	34	313
12	I 68-70	23	69-72	57	90-97, 124

Grand Rapid to Cumberland House

It may be recalled that this section of the route, which both Companies used, was omitted from the last chapter. Like Lake Winnipeg, it is neither attractive nor distinguished. Grand Rapid now lies beneath a headpond, which floods out also the rapids which caused the Portage of the Roche Rouge, the Grand Décharge, and the several rapids just below Cedar Lake; Cedar Lake is converted to a scene of man-made, instead of natural, swampy bleakness. What has gone is described well and fully in Elliott Coues (ed.) *New Light . . . II*, pp. 461-66. From the delta leading into Cedar Lake the voyageurs fought the fast current of the Saskatchewan. Other difficulties were the constantly shifting mud banks and the absence, sometimes for miles, of camping or even landing spots on the swampy shores. Except near The Pas, the Saskatchewan is featureless all the way to Cumberland House — where the Montrealers left the Saskatchewan and headed north.

The Sturgeon-weir River

The Sturgeon-weir is a tributary of the Saskatchewan, connecting this river with the Churchill River. To the voyageurs, who remembered only going up its rapids, it was the *Rivière Maligne*. Coming *down* it was sheer joy. The Sturgeon-weir is not recommended as an up-route, since the river's rate of drop (at four feet to the mile) is a nearly ideal gradient to come down. The diary for a trip I made thirteen years ago down the Sturgeon-weir claims that we portaged a dozen times; in my memory, however, the river stays, in Mackenzie's words, "an almost continual rapid". The dozen portages listed by Mackenzie — the Pente, Bouleau, de l'Isle, d'Epinettes, three *galets*, to name the principal ones — are still in use; and the modern voyageur is back again in an earlier century with his predecessors. Another delight, after Lake Winnipeg and the Saskatchewan, is the clear water. The Shield is re-entered at Amisk Lake, a little way up from Cumberland. At Frog Portage, an easy three-hundred-yard portage leads from a stagnant lagoon, the source of the Sturgeon-weir, over to the main stream of the Churchill in Trade Lake.

The Churchill River

The Churchill, in the journals of its early travellers and in the memory of those who have more recently travelled it, stands out as one of the best canoe rivers on the whole route. It is less a river than a chain of attractive lakes, mostly in rugged rock country. Wherever it narrows down there is a rapid or a fall, as one lake spills into the next. In the five hundred miles between Ile-à-la-Crosse on the Churchill and Cumberland House at the mouth of the Sturgeon-weir, there are about a hundred rapids, of which sixty-five can be run. There is road access to Churchill water at Flin Flon, Man., and from Prince Albert via either Lac la Ronge or Green Lake. Air access is from Ile-à-la-Crosse or Flin Flon. Anyone planning this trip is advised to read Sigurd Olson's splendid *"The Lonely Land"*, which brings out the full richness of the Churchill's history and wilderness, as well as the excitement of paddling this great river. The country is unchanged; the

Churchill is still the only highway through. In Dipper Lake, two below Ile-à-la-Crosse, the traveller up the Churchill leaves the Shield again, and the country above loses some of its scenic character.

From Ile-à-la-Crosse Post the Athabasca Brigade headed over the open horizon of the Lake's Aubichon Arm. They skirted the south shore of Churchill Lake, passed Buffalo Narrows, and crossed Peter Pond Lake. Now came toil once more, as they headed up the current of the swift, shallow Methye (La Loche) River; frequently they had to get out to wade or line; many ledges called for poling. A single swamp's snaking loops took us eight hours' upstream paddling. At the head of the Methye River lies Lake La Loche. Now the voyageur was approaching a portage that he would be telling his grandson about.

Methye Portage to Fort Chipewyan

The marathon, twelve-mile Methye Portage starts at a point a mile up a small winding creek from Lake La Loche. Writers, who obviously have not seen Methye Portage, have described it as hilly, which is not the case, except for the well-advertised big drop at its north end. The portage crosses a low sandy ridge in sparse spruce and jack pine; it deviates but little, and heads for a small lake eight miles away. This lake, a mile across, gives brief relief from the bite of the pack strap. Bordered by beaches of white sand, it offers a pleasant way to wash off some of the sweat. Where the trail comes in, is a spacious camp-site, level and parklike, and also historic. This was "The Rendezvous" of the Hudson's Bay Company La Loche and Mackenzie Brigades in the peak days of the Portage.

Here is a fitting context for a digression on the ultimate Hudson's Bay Company logistics, which closely paralleled those of the North West Company before Union. The object was to get furs from the vast Mackenzie District on board sea-going vessels at York Factory, and the trade goods back — impossible for a return run within the ice-free months. The solution was a neat one. Furs from everywhere west of Lake Superior were collected at Norway House and taken down by a large brigade of York Boats in August in time to catch the ship at York Factory. This brigade neither wintered there nor returned with empty boats: the new shipment of trade goods was picked up and carried to Norway House, to be stored and sorted over winter. Taking the Fort Garry shipment only, the brigade then went south on Lake Winnipeg to winter around the Forks of the

Red and Assiniboine. In early June, by which time Lake Winnipeg was free of ice, the La Loche Brigade started off with nearly empty York Boats, for Norway House. The whole of the year's "Outfit" for the Mackenzie, dropped off the previous autumn, was picked up and transported by the familiar Cumberland-Churchill River route all the way to Methye Portage. The boats were left at the portage landing, but the tons of supplies were transported eight miles to The Rendezvous at the little lake, here to be picked up by the Mackenzie Brigade, whose boats were waiting at the north end of the Methye Portage. No voyageur returned unloaded over the portage, for the previous year's collection of Mackenzie furs had to be carried east to York Factory.

Methye Portage thus performed the function of the North West Company's Grand Portage, avoiding the need to portage the boats themselves; Norway House matched the function of Rainy Lake as an advance depot, making possible a reduced return journey within the season. York Factory, of course, was the Hudson's Bay Company's "Lachine". Methye Portage till 1885 was the principal link of any kind between the Hudson Bay and Mackenzie River drainages. Since these two basins between them cover nearly two-thirds of Canada, its freight and passenger traffic were once considerable. In its latter days the portage became "mechanized", through use of horses and oxen.

When I crossed the Methye Portage in 1958, I was interested to note, in the extensive clearings and flattened building-sites at both ends, the evidence of its earlier heavy use. A dry climate and sandy soil make it unlikely that it will become grown over. The trail is wide enough still for a waggon. An old Indian we spoke to recalled hearing his grandfather talk of the once busy traffic.

The portage trail climbs gently from the "Rendezvous" lake's opposite corner and heads for the Clearwater, four miles away. Soon, without warning, the land falls suddenly away to a view thirty or forty miles down the Clearwater valley. Mackenzie, Franklin and others went into such raptures over this view that it is a little disappointing by contrast; all that is offered is a distant prospect down a narrow valley, from a seven-hundred-foot drop. The portage path drops so steeply from this point that loads and oxen were shifted from carts to crude sledges to be dragged down to the Clearwater and the waiting York Boats.

The "Clearwater" River, as so often happens with geographical names, indicates less a presence than a neighbouring absence: the indicated characteristic is merely relative. The Clearwater is clearer, and the view down its valley more attractive, for the proximity to uninteresting country to the south. Like the Athabasca, it is a fast river, and anyone planning to paddle between Ile-à-la-Crosse and Fort Chipewyan (not specially recommended) should certainly do it from south to north; in the eighty miles of the Clearwater are a fall and six rapids. The Clearwater is also noteworthy for being the only west-flowing river on the route west of Lake Winnipeg.

The 200-mile run down the Athabasca River from Waterways is fast but dull. Approaching Lake Athabasca, a smaller outlet breaks off, called the Embarras River, which the voyageurs usually followed. "Embarras" was a voyageur term for debris, mostly tangled driftwood and uprooted trees. It would appear that "embarras" often blocked off the start of this outlet, for there is occasional reference to a portage or lift-over. The Embarras "River" offered a more direct route to and from Lake Athabasca, and — most important of all — elimination of much open-lake paddling on this unfriendly water.

Fort Chipewyan deserves special attention for its place as the terminus of a trade route. Why at Fort Chipewyan? — not at Ile-à-la-Crosse, or Fort St. John, or New Caledonia? Fort Chipewyan, whether at its first south-shore site, or at its later location, uniquely fulfilled a variety of critical requirements. In the first place, it was about as far west as was possible to allow canoes to meet the eastern brigades and still get home before the ice. How marginal this distance was will appear in the next chapter. In the second place, Fort Chipewyan was close to the mouth of the Athabasca. A fast river such as the Athabasca breaks up a month before a large lake, like Athabasca. Specifically, their respective break-up dates were about May 15 and June 15. The month's interval spelled the difference between getting back in time from Rainy Lake or not. Thirdly, a key depot such as Fort Chipewyan required sufficient staff to cause a problem in provisioning, while Lake Athabasca provided whitefish in abundance, as well as occasional moose and other game. What multiplied its importance was the tremendous area beyond it, including the basins of the Mackenzie, the Liard and the Peace, from which its furs were drawn.

Cumberland Ho. to Ft. Chipewyan Reading

| 9 | II 473-76 | 18 | 138-40 | 27 | 79-96 |
| 12 | I 278-98 | 20 | 261-69 | 32 | 34-38 |

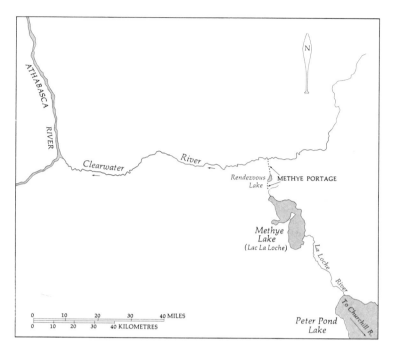

Methye Portage

Ancillary and Connecting Routes

Ancillary and connecting routes in the area west of the Great Lakes related almost entirely to the Albany and Churchill Rivers. A well-known canoe route via the English and Albany Rivers links James Bay with the Winnipeg River. The portage over the height of land lies between Lac Seul and Lake St. Joseph. There is record of a Hudson's Bay Company party having got through via the Albany, English, and Winnipeg Rivers from James Bay all the way west to Brandon. In considering the Albany-English route as a regular freight route to the Prairies, however, one needs merely to observe that the crest of the route at Lake St. Joseph lies at over 1200 feet, whereas via the Hayes River the rise is only to Lake Winnipeg, at 700 feet. This route today suffers at one end from too many dams, and (owing to diversion) at the other from too little water.

The Churchill River had canoe connections (apart from those already noted) with:

(a) The Nelson River, via the Burntwood River;
(b) Lake Athabasca, via Reindeer Lake and the Fond-du-Lac River;
(c) The Athabasca River, via the Beaver and Lac la Biche.

Two principal canoe routes led from York Factory to Frog Portage on the Churchill River: the main one already described via Norway House and Cumberland House, and a route lying in an almost direct line, by way of the Nelson and Burntwood Rivers and a portage over to the Churchill River via the Kississing. The latter route's directness, taking one side of a triangle, appealed to George Simpson and he himself used it on his journey of 1824. Eleven years later, however, Chief Trader Donald Ross reported that the Nelson-Burntwood route had been abandoned, and the "common route to and from the factory" was still being used. The reasons are various. In the first place, there was serious risk (which Simpson himself experienced) entailed in an open passage on Hudson Bay around Point of Marsh to start up the Nelson from York Factory. Second, was the firm reluctance of the voyageurs to forfeit the passing delights of Norway House and Cumberland en route. Third, and most interesting, was the almost paradoxical discovery that the much shorter route was more expensive; Nicholas Garry in weighing the various considerations stated: "Depots of provisions must be brought from Cumberland House at Split Lake, Burnt-wood Portage or Nelson House and the Frog Portage, whereas in the present (Cumberland) Route . . . the supplies are already in the Way, and it will require considerable Expense to convey the Provisions . . . to the Nelson River Route." This illustrates how the Prairies region, though actually non-productive of furs, played its part in supplying fuel for transport; and also why the fur-trade trunk route had to flank the Prairies, in order to facilitate delivery to the pemmican posts of Ile-à-la-Crosse, Cumberland, Norway, and Bas-de-la-Rivière. The Nelson, as mentioned earlier, is not good canoeing; the Burntwood has some fine waterfalls, but is not otherwise specially attractive; nor is its water clear; its lower reaches, moreover, may become affected by the planned Rat River diversion.

A route explored by David Thompson in 1796 went up Reindeer Lake, thence directly over to Wollaston Lake and down the Fond-du-Lac River. Though this route would have been a shortcut and back door to the Athabasca Country, besides having the advantage of bypassing the Methye Portage and River, it proved an abortive attempt. Its two big lakes do not break up in time for a fast spring getaway. Though it thus failed to become a trade route, it is an excellent modern canoe trip, very scenic and exciting.

Finally, a connecting route with the Churchill River went up from Lake Ile-à-la-Crosse by the Beaver River, over a bad divide, and down the La Biche to the Athabasca River. Simpson, heading up the Athabasca to cross the Rockies in 1824, chose this route — influenced again, as in his mistaken judgment of the Burntwood, by its mere directness. After he had seen it, he wrote of its "tediousness and uncertainty . . . even to the half-laden North Canoes, rendering the Voyage extremely harrassing to the people . . . and in dry Seasons incurring the risk of not getting to the Wintering Grounds before the Ice sets in". George Simpson seems to have been the last to take the Beaver-La Biche route. It was too meandering and shallow to serve as a trade route. Nor should anyone think seriously of it as a canoe trip today.

Chapter VII

Crossing the Rocky Mountains

In a story of fur trade canoe routes, this chapter could be entitled "An Attempt That Failed". Strictly, over the Rockies was a non-canoe route, for the problem of servicing the trading posts beyond the mountains was ultimately solved by a combination of sailing vessels, bateaux and horses. The two trade termini, Edmonton and Fort Chipewyan, became connected with the Pacific, not by a trade route, but an express route. The struggle to cross the Rockies by freight canoes makes an intriguing story. Reference to fur trade routes in the Mackenzie basin and Yukon Territory will also be included in this chapter.

Two main fur areas lay beyond the Rockies: (a) the *Columbia Department*, comprising the lower Columbia Valley, and based on both David Thompson's explorations and the taking over of Astor's Pacific Fur Company by the North West Co. in 1813; and (b) *New Caledonia*, lying north of the bend of the Fraser, and associated with the explorations of Mackenzie and Fraser. The problems in servicing these two areas were steep gradients, early snow, late spring, and distance.

Access to the Columbia Department was by way of the sea at the mouth of the river, or by one or another pass leading over the mountains from either the Saskatchewan or Athabasca River. Access to New Caledonia by water was by way of the Peace River.

The early attempts to set up a trade route feasible for freight were discouraging. Though Fraser in 1808 achieved the incredible feat of descending his river, even he had to abandon his light canoes for one portion of the river and secure new ones later. Mackenzie in 1793 had had to abandon the Fraser near Quesnel and finish on foot. David Thompson had been turned aside on the North Saskatchewan at Rocky Mountain House by the Piegans, jealous of their transmontane enemies receiving arms. Thompson by 1807 had succeeded in reaching the Columbia via the steep Blaeberry River, but on horseback, and as a miracle of survival.

New Caledonia was simpler than the Columbia to reach from the East: Fort McLeod, at 2200 feet altitude, was only 1500 feet above Fort Chipewyan; no serious portaging except the dozen miles at Hudson Hope lay

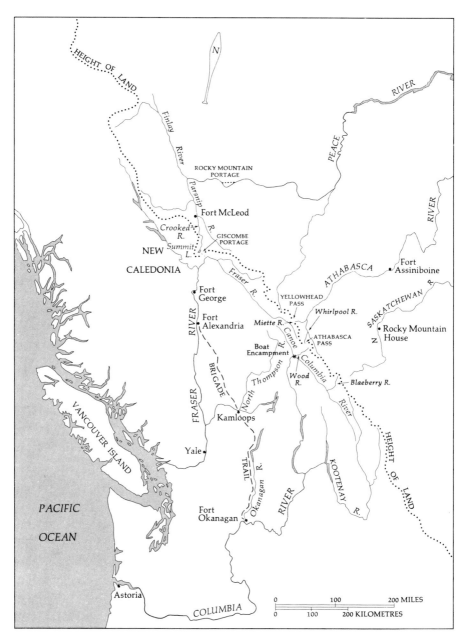

Crossing the Rockies

between. The problem was distance — nearly a thousand miles — and distance had to be weighed in terms of frost. The trade route westward was already over-extended at Lake Athabasca. A brigade returning from Rainy Lake did not reach Fort Chipewyan usually till late September, a precariously late start for re-forwarding the New Caledonia outfit. Daniel Harmon records in his diary for November 7, 1818:

> "We have now about a foot of snow on the ground. Today our people returned from the Rainy Lake, and say that, on account of the large quantities of ice that was drifting in the Peace River, they were obliged to leave the greater part of the goods . . . but a short distance this side of the Rocky Mountain Portage. We shall be obliged, therefore, to bring these goods on sledges, drawn by dogs from that place, which is distant from this, about 280 miles."

Nor could the furs get through to Rainy Lake in the same season, because ice locked the Parsnip and Peace for as late as it did the Athabasca River, preventing the winter's furs reaching Fort Chipewyan in time to catch the brigade, which left around May 15.

The Peace was abandoned as a main freight route in 1826 in favour of a solution described below. Mainly because of the long portage at Hudson Hope (the present dam site), North Canoes, not York Boats, were used on the Peace. To avoid a more roundabout water trip, the canoes were unloaded at Fort McLeod, and the Stuart Lake consignment was taken over the divide by trail. The attractiveness of retracing this route today has been reduced by the construction of the Peace River Dam. The Peace River has only one other obstruction, a drop over a low limestone shelf at Vermilion. This river, like the Mackenzie, displays its grandeur better from on top of its banks than at water level.

The Columbia Department, posing the much more difficult problem to service, strictly belongs outside the scope of a study of fur-trade canoe routes: only the *approaches* to the Rocky Mountains, on both sides, were ever tried by canoe — and then only briefly. The freight route first used for the Columbia Department followed the Athabasca River by canoe westward to about Jasper, thence by pack horses up the Whirlpool River and over 5700-foot Athabasca Pass, past a little tarn, the "Committee's Punch Bowl", sitting on the divide. The horses then dropped steeply 3500 feet to the Columbia River, where Boat Encampment depot stood, between the mouths of the Wood and Canoe River tributaries. A new set of North Canoes — later bateaux — waited at Boat Encampment to take the goods to the sea, dropping still another 2000 feet, with several portages. It was the return journey in late April which killed this route — first breasting the spring current of the Columbia, then the 3500-foot climb from Boat Encampment over the Pass, not on horseback, but on snowshoes, carrying heavy packs. Simpson, after doing the trip himself in 1824, wrote it off as a freight route. It continued to be used as the express route for the Columbia for some years. To speed communication between the Athabasca and the Saskatchewan, Simpson had a horse trail cut from Fort Assiniboine on the Athabasca to Edmonton.

Of the once canoeable portions of this route, the Columbia today is on the way to becoming no longer a canoeing river, owing to a procession of new dams. Paul Kane's journal for November 16 to 29, 1846, gives a vivid account of the twelve-day descent of the Columbia by bateau ("formed canoe fashion, with round bottoms of boards, clinker built").

The ultimate solution to "crossing" the Rockies with freight was to go around them, involving a long and hazardous detour by sea from London around Cape Horn to the mouth of the Columbia. This trip was initiated in 1813 by the *Isaac Todd*, and regularized after 1826. In 1813, too, came the significant discovery by John Stuart of a pack trail leading from the foot of navigation on the upper Fraser at Fort Alexandria over to Kamloops, thence passing along the west side of Okanagan Lake to Fort Okanagan, on the Columbia near the present border. Though for a short time after 1813 New Caledonia continued to ship its furs a year late, via the Peace and Fort Chipewyan by canoe, from 1826 the horse brigades on Stuart's route took over the complete servicing of New Caledonia. The brigade route was shifted in 1847, after the Oregon crisis, to pass directly from Kamloops to the Fraser; and eventually it was shifted after 1862 to the Cariboo Road.

The Fraser River below Fort Alexandria (near Quesnel) was at no time a fur-trade canoe route. George Simpson required convincing; but after a trip which he personally made down the Fraser with an expert crew in 1828, he wrote:

> "*Frazers River . . . was never wholly passed by water before, and in all probability never will again . . . and altho we ran all the rapids in safety, being perfectly light and having three of the most skilful Bowsmen in the country, . . . I should consider the passage down to be certain Death, in nine attempts out of Ten. I shall therefore no longer talk of it as a navigable stream.*"

"The Committee's Punch Bowl", a small tarn sitting on Athabasca Pass. From one end the water flows into the Whirlpool River and eventually, the Mackenzie; from the other end of the tarn, water flows out to start the Wood River, emptying at Boat Encampment into the Columbia River.

A drawing by C. W. Jefferys, "Simon Fraser descending the Fraser River, 1808"
It shows North Canoes lightly laden, with small crews for maximum freeboard
in the turbulent Fraser canyon.

One further connecting (but not canoe) route beyond Edmonton was *down* the Athabasca River to Waterways, and thus bypassing the rugged Methye Portage route. Horses took the goods north from Edmonton by road ninety miles to Athabasca Landing. Barges were then floated down the fast Athabasca, their cargo being portaged only once on the route, over the island at Grand Rapids. They would be lined up on the return journey. This barge route operated only between 1885 and the building of the railway from Edmonton to Waterways in 1917.

Canoe travel on the Athabasca still is attractive and unspoiled. The modern voyageur on the trip from Jasper to Waterways should, however, remember that the Athabasca River, as a mountain stream, fluctuates sharply in volume, depending on the temperature near its source; a dry spell happening to coincide with cool weather can so reduce its volume that river-broadenings such as Brule Lake become unnavigable, shallow, bouldery areas — and without advance notice.

Crossing the Rockies Reading

9	II 642-62, 675-96, 747-916	33	31-37, 76, 265, 349-50
13 & 27	(general)	34	293, 344-87
18	144-59, 191-204, 216-33	51 & 53	(general)
23	99-116, 225-55	57	70-86, 103-20
29	9-13		

The Mackenzie and the Yukon

Only one large area has so far been ignored in this volume: the Mackenzie and Yukon River basins. The limitation which a five-months' navigation season placed on freight-canoe travel precluded a trunk route this far north. The Mackenzie, the Peace, and the Liard simply served as giant feeders for the *entrepôt* at Fort Chipewyan.

The general appearance of the country down the Mackenzie River has not changed greatly in its two centuries of fur trade; its navigation, with the one sixteen-mile break between Fitzgerald and Fort Smith, has changed not at all. Down the Mackenzie's swift current, fifty miles a day is about par for canoe travel. However, for the benefit of the surprising number of inquirers who seem romantically attracted to retrace Mackenzie down this Mississippi of the North, it should be stressed that the Mackenzie is not really worthy of their interest: its scenery lies too far back; its banks are

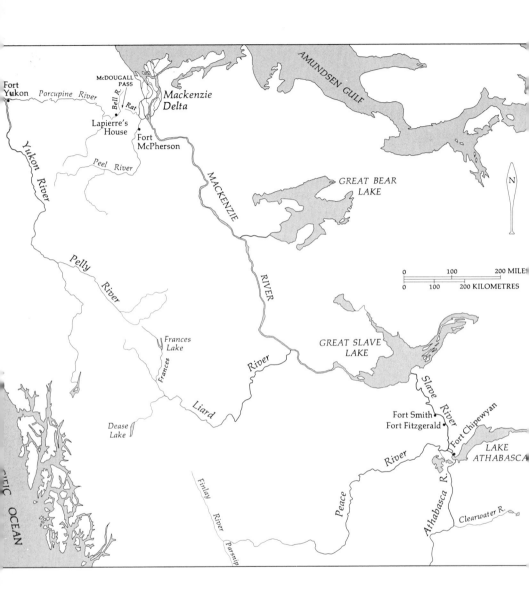

The Mackenzie and Yukon connections

of clay, and its water, murky; barge traffic and oil drums mar the atmosphere.

The furs from the Liard River posts were brought down to Fort Simpson, at the junction with the Mackenzie. In penetrating the Liard basin, a connection (via Dease Lake) even with the Stikine River flowing into the Pacific was found, but later abandoned. Though the furs were both good in quality and plentiful, the Liard as a freight canoe route posed problems, in its main canyon. This river, like the Peace, rises west of the Rockies, and therefore has to pierce them. A contemporary description by George Keith in a letter to Roderic McKenzie in 1807 said of the Liard:

> "About one and a half days march from Mackenzie's river, it becomes a continuation of rapids for upwards five leagues, and so shallow ... that some years, in the Fall, a canoe loaded with six or eight pièces, cannot proceed without making an almost continual portage ... In spring and in the beginning of summer there are none, but the navigation is very difficult on account of the strength of the current. The banks all along and on each side of these rapids, are very high and perpendicular, consisting of broken rocks."

The Liard Canyon's difficulties are cited here as background for another extraordinary logistical solution — such as already noted for the Columbia, by ship around Cape Horn. Robert Campbell of the Hudson's Bay Co. in 1851 ascended a tributary of the Liard, the Frances River, which opened up a whole new source of furs. From Frances Lake he found a route (via Finlayson and Campbell Rivers) over to the Pelly River, which he followed down, confirming his surmise that it was functionally the Yukon. At Fort Yukon, Campbell turned eastward and went up the Porcupine River and its tributary, the Bell River. From a point high on the Bell, an Indian trail led over the Richardson Mountains to Fort McPherson, near the junction of the Peel and Mackenzie Rivers. The Hudson's Bay Co. established Lapierre's House as a small depot where this trail joined the Bell, and thenceforward used this very circuitous route to service all the posts set up in the Yukon and Pelly regions.

The topographical reason for this surprising detour (see map, p. 106) was not only the Liard's difficulties, but the extraordinary accessibility of the continental divide near Lapierre's House. The nameless pass on the dogsled trail over to Fort McPherson stood at only 2000 feet above sea level. This winter-trail followed two mountain creeks whose source streams almost touched in the pass. The total distance from Lapierre's House to Fort McPherson was about seventy miles.

Many Klondikers who chose the route via Edmonton and the Mackenzie River passed over the mountains by way of a water route which flanked the winter trail. The Rat and Porcupine Rivers rose in the area of McDougall Pass a little to the north, which (at only 1000 feet) was by far the lowest nick in the whole 1700-mile mountain rampart stretching from the U.S. boundary to the Arctic Sea.

If anyone is looking for the most physically challenging section of old fur-trade routes to retrace today, the Rat-Porcupine trip over the mountains should satisfy him. Starting at Fort McPherson just above the Arctic Circle, instead of following the parallel trail, it is possible to go up the Rat River by canoe, first surrendering any idea of paddling. Nine days' lining, poling, and wading in icy water (averaging five miles travel a day) bring one to the dramatically beautiful pass, where the barren Richardsons rise starkly in peaks 4000 feet above the river. Then a single half-mile portage (the only one in the 1500 miles between the mouths of the Mackenzie and Yukon Rivers) puts the canoes on the Little Bell, which empties eventually into Bering Sea. After the first day's downstream travel, the earlier slow ascent is more than offset by a standard speed of fifty or sixty miles a day on the fast Bell and Porcupine Rivers. In 1965, Lapierre's House, now just a group of abandoned wooden buildings, still stood. Another mountain-river trip in the Yukon Territory is the portion of Robert Campbell's route from Frances Lake over to the Pelly River and down to Fort Yukon. The latter is a less challenging route over the divide, since its start is so close to the summit — and Finlayson Lake in the spring sometimes flows out both ways — but the route down is much faster, starting from 3000 feet higher above the same terminus. The Pelly River route is described well in Warburton Pike's entertaining *"Through The Sub-Arctic Forest"* (1896).

Conclusion

In a decade when some Canadians are vying with one another to discover and exploit their differences, it is healthful to review a story in which French and English, Indians, Métis, and Scots all worked closely together. In the final stages, after 1821, there ceased to be even the bitter rivalry between the Montrealers and the Hudson's Bay Co. The fur trade as a whole, from wherever run, was a vast Canadian enterprise.

The fur trade was a decisive factor in determining Canadian boundaries — principally just west of Lake Superior, and in British Columbia. Canada's extension from sea to sea, in the face of American "Manifest Destiny", is attributable largely to the fur trade. In Harold Innis's words, "It is no mere accident that the present Dominion coincides roughly with the fur-trading areas of northern North America".

The fur trade was responsible for the ascendancy of east-west thinking over strong geographical and economic north-south influences. Thirty years before Americans had a foothold even west of the Mississippi, there was in Canada the steady pulse of east-west commerce between Montreal and the Athabasca Country. In such developments, and more particularly in the penetration by Canadien voyageurs and *coureurs de bois* into the Northwest and the Rocky Mountains, were the geographical foundations of Canada securely laid.

Practically every important Canadian explorer, as distinct from discoverer — Champlain, Mackenzie, La Vérendrye, Brûlé, Nicolet, Thompson, Fidler, Hearne, Turnor, Fraser, Kelsey, Henday, Radisson, Groseilliers, and Campbell — was either the servant of a fur company or in some other way connected with the search for furs and fur routes. The mystic "Search for the Western Sea" played only an initial and minor part in the exploration of Canada — which was almost totally a function of the fur trade. The beaver by its defencelessness, no less than by its value, was responsible for unrolling the map of Canada.

For "do-it-yourself" historians who may feel the urge to pursue this significant phase of Canadian history, it must be obvious from these pages that mere reading can give only a flat, two-dimensional picture. To take copies of the early journals out into the field for study on the spot, in an environment totally unchanged as to vegetation, water currents, even exact pathways, gives depth and life to the record. To realize that so much still lies unchanged is the first step; this is but a legacy of our wilderness. The next step is to be persuaded that it can be done, and done easily, by anyone experienced in handling a canoe. This is the only craft suitable for extended travel along wild and interrupted navigation. Questions of logistics are raised, the doorway to whose solution lies through maps. Map distribution is handled by the Department of Energy, Mines and Resources; the address is 615 Booth St., Ottawa. On application, this office will send a key map indicating the map numbers for different areas, and their scale, from which to order.

Matters of food and equipment are easily arranged in southern areas close to home. For the more ambitious who would explore farther north,* enquiries should be addressed to Northern Stores Department, Hudson's Bay Co., Winnipeg 1. This three-centuries-old company is so forward-looking that it has applied car-rental practices to canoes; under their "U-Paddle" service, canoes can be picked up at one Hudson's Bay Company post and left at another post at the end of a journey. As to routes, it has been my experience that any canoe route once plied regularly by the fur canoes is still good canoeing today.

"No less than ten brigades, each numbering twenty canoes, used to pass through these scenes during the summer months . . . I have seen four canoes sweep round a promontory suddenly and burst upon my view, while at the same moment the wild romantic song of the voyageurs, as they plied their brisk paddles struck upon my ear; and I have felt thrilling enthusiasm on witnessing 30 or 40 of these picturesque canoes . . . half shrouded in the spray that flew from the bright vermilion paddles.
"Alas! the forests no longer echo to such sounds . . . and nought save narrow pathways over the portages, and rough wooden crosses over the graves of the travellers who perished by the way, remains to mark that such things were." — R. M. Ballantyne: Hudson Bay.

While the lively and colourful scenes which once animated the old fur trade highways are no more, their "then" and "now" are similar enough that these canoe routes will long remain to attract the historically minded and to challenge the adventurous.

*Arctic canoeing requirements are discussed in my Summer Travel In The Canadian Barren Lands (Can. Geog. Journal, May 1967).

Suggested Voyageur Reading

I Primary Printed Sources:

(Most of this list are out of print, but possibly available in reference libraries or in reprints.)

1. Ballantyne, R. M.: *Hudson Bay*. London, 1879.
2. Belcourt, G. A.: *Mon Itinéraire*. (Bulletin de la Société Historique de Saint Boniface, IV, 1913).
3. Bigsby, J. J.: *The Shoe and the Canoe*. London, 1850. 2 vols.
4. Butler, W. F.: *The Great Lone Land*. London, 1875.
5. Butler, W. F.: *The Wild North Land*. Montreal, 1874.
6. Campbell, Robt.: *Discovery and Exploration of the Youcon*. Winnipeg, 1885.
7. Champlain, S. de: *Voyages and Explorations*. Toronto, 1911. 2 vols.
8. Cocking, M.: *Cumberland House Journal, 1775-6*. (in Cumberland and Hudson Ho. Journals, H. B. Record Society).
9. Coues, Elliot, (ed.) *New Light on the Early History of the Greater Northwest*. New York, 1897. 3 vols.
10. Delafield, Maj. J.: *The Unfortified Boundary*. New York, 1943.
11. Fountain, Paul: *The Great North-West and Great Lake Region of North America*. London, 1904.
12. Franklin, (Sir) J.: *Narrative . . . 1819-22*. London, 1824.
13. Fraser, Simon: *Letters and Journals*. Toronto, 1960.
14. Garry, Nicholas: *Diary* (Transactions of Royal Society of Canada, 1900).
15. Gates, C. M. (ed.): *Five Fur Traders of the Northwest*. St. Paul, 1965.
16. Grant, G. M.: *Ocean to Ocean*. London and Toronto, 1873.
17. Hadfield, Joseph: *An Englishman In America 1785*. Toronto, 1933.
18. Harmon, Daniel: *Journal of Voyages and Travels.*.(Trail Makers of Canada Series) Toronto, 1911.
19. Hearne, S. and Turnor, P.: *Journals*. (ed.) J. B. Tyrrell, Champlain Society, Toronto, 1934.
20. Henry, Alexander: *Travels and Adventures*. (ed.) J. Bain, Toronto, 1901.

21. Heriot, George: *Travels Through the Canadas.* London, 1807.
22. Hind, H. Y.: *Narrative of the Canadian Red River Exploring Expedition.* London, 1860. 2 vols.
22A. Hood, Robert: *To the Arctic by Canoe, 1819-21.* Montreal, 1974.
23. Kane, Paul: *Wanderings of an Artist.* Toronto, Radisson Society, 1925.
24. Keating, W. H.: *Narrative of an Expedition.* Philadelphia, 1824. 2 vols.
25. Kelsey, Henry: *The Kelsey Papers.* Ottawa, 1929.
26. Lefroy, J. H.: *In Search of the Magnetic North.* Toronto, 1955.
27. Mackenzie, Alexander: *Voyages From Montreal.* Toronto, Radisson Society, 1927.
28. Masson, L. R.: *Les Bourgeois de la Compagnie du Nord-Ouest.* (Reprinted) New York, 1960. 2 vols.
29. M'Gillivray, Duncan: *Journal.* Toronto, 1929.
29A. The Papers of the Palliser Expedition, ed. I. Spry, Champlain Society, Toronto, 1968.
30. Pike, Warburton: *Through the Sub-Arctic Forest.* London, 1896.
31. Radisson, Pierre Esprit: *Voyages.* (Reprinted) Minneapolis, 1961.
32. Simpson, George: *Athabasca Journal, 1820-21.* H. B. Record Society.
33. Simpson, George: *Journal, 1824-25.* [Fur Trade and Empire (ed.) F. Merk] Cambridge, Mass., 1931.
34. Thompson, David: *Narrative.* (ed.) R. Glover, Champlain Society, Toronto, 1962.
35. Tyrrell, J. W.: *Across the Sub-Arctics of Canada.* Toronto, 1908.

II Secondary Works:

36. Burpee, L. J.: *The Search For The Western Sea.* Toronto, 1908. 2 vols.
37. Campbell, Marjorie W.: *The North West Company.* Toronto, 1957.
38. Crouse, N. M.: *La Vérendrye.* Toronto, 1956.
39. Dawson, S. E.: *The Saint Lawrence Basin.* London, 1905.
40. Glazebrook, G. de T.: *A History of Transportation in Canada.* Toronto, 1938.
41. Innis, H. A.: *The Fur Trade In Canada.* New Haven, 1930.
42. Martin, H. T.: *Castorologia.* London, 1892.
43. Morton, A. S.: *A History of the Canadian West to 1871.* London.
44. Nute, G. L.: *The Voyageur.* New York, 1931.
45. Nute, G. L.: *The Voyageurs' Highway.* St. Paul, 1941.
46. Patterson, R. M.: *Trail To The Interior.* Toronto, 1966.

47. Rich, E. E.: *Hudson's Bay Company, 1670–1870*. Toronto, 1960. 3 vols.
48. Rich, E. E.: *The Fur Trade and the Northwest to 1857*. Toronto, 1967.
49. Wallace, W. S.: *The Pedlars From Quebec*. Toronto, 1954.

III Monographs, Articles:

50. Alcock, F. J.: *Past and Present Trade Routes*, in "The Geographical Review", Vol. X, 1920, pp. 57-83.

    In *The Beaver*, H. B. Co.
51. Creech, E. P.: *Brigade Trails of B.C.* March, 1953.
52. Heming, Arthur: *The Abitibi Fur Brigade*. Summer, 1967.
53. Patterson, R. M.: *We Clomb The Pathless Pass*. Winter, 1960.
54. Simpson, Frances: *Journal*. Dec. 1953; Mar. & Summer, 1954.
55. Turner, J. P.: *The La Loche Brigade*. Dec., 1943.
56. Buck, Solon J.: *The Story of Grand Portage*. (16pp.) Minneapolis, 1931.
57. Ermatinger, E.: *York Factory Express Journal*. (Transactions of the Royal Society of Canada, 1912).
58. Glover, Richard: *Difficulties of the H. B. Co.* C. Hist. Review, pp. 240-54, Sept. 1948.
59. Umfreville, Edward: *Nipigon to Winnipeg*. (Journal) 63 pp. Ottawa, 1929.
60. Wheeler, A. O.: *Autumn On The Upper Ottawa*. The Ottawa Naturalist, pp. 81-93, 1888.

The National Historic Sites Service is a part of the National and Historic Parks Branch of the Department of Indian Affairs and Northern Development. It is responsible for the acquisition, restoration, interpretation and maintenance of national historic parks and sites to commemorate persons, places and events of national significance in the history of Canada. It presently maintains or is in the process of restoring and developing some 35 parks and major sites from coast to coast.

The biggest single project undertaken by the Service to date is the restoration of the Fortress of Louisbourg which will become a fully operational historic park in 1972.

Among the historical themes already or shortly to be covered by the Service in its parks and sites are the fur trade, early settlement, coastal defence, science, political developments, the opening of the west, Anglo-French rivalry, Canadian-U.S. relations, the Gold Rush of 1898 and Indian history and culture.

In addition to active development and maintenance of parks and sites, the Branch is launching a comprehensive new publications program ranging from brief colourful handout folders on each park to scholarly treatises upon various archaeological topics, and designed to appeal to many different groups of readers.

Included in the program is a series of books dealing with major historic themes, of which this is the first. It stems from a recommendation made early in 1967 by the Fur Trade Committee of the Historic Sites and Monuments Board of Canada, the statutory board appointed to advise the Minister of Indian Affairs and Northern Development on all matters of national historic importance. Its members are:

Newfoundland —
Dean Leslie Harris, M.A., Ph.D.

Nova Scotia —
Professor Peter B. Waite,
M.A., Ph.D.

Prince Edward Island —
Reverend Francis W. P. Bolger,
S.T.L., Ph.D.

New Brunswick —
George MacBeath, Esq.,
B.A., D. de l'U. (Paris)

Quebec —
M. Jean-Jacques Lefebvre,
B.L., F.R.S.C.,
Chief Archivist

Professor Marc La Terreur

Ontario —
Professor Donald G. Creighton,
C.C., M.A., D.Litt., LL.D., F.R.S.C.

James J. Talman, Esq., C.D., M.A.,
Ph.D., D.Litt., F.R.S.C.

Manitoba —
George Anderson, Esq.

Saskatchewan —
Allan R. Turner, Esq., M.A.
(Chairman of the Board)

Alberta —
Lewis H. Thomas, Esq.,
M.A., Ph.D.

British Columbia —
James K. Nesbitt, Esq.

Ex officio —
Wilfrid I. Smith, Esq., M.A., Ph.D.,
Dominion Archivist

W. E. Taylor, Esq., A.M., Ph.D., F.R.A.I.,
Director, National Museum of Man

Peter H. Bennett, Esq., M.A.,
(Secretary),
Assistant Director (Historic Sites),
National and Historic Parks Branch.

John I. Nicol,
Director,
National and Historic Parks Branch,
Ottawa, Ontario.
June 1971.

Index